C. Wagener S. Neumann (Eds.)

Molecular Diagnostics of Cancer

With 50 Figures and 31 Tables

Springer-Verlag
Berlin Heidelberg New York
London Paris Tokyo
Hong Kong Barcelona
Budapest

Professor Dr. Christoph Wagener
Abteilung für Klinische Chemie, Medizinische Klinik
Universitäts-Krankenhaus Eppendorf
Martinistraße 52, W-2000 Hamburg 20, FRG

Dr. Siegfried Neumann
Diagnostische Forschung E. Merck
Frankfurter Straße 250, W-6100 Darmstadt, FRG

ISBN-13: 978-3-540-55476-9   e-ISBN-13: 978-3-642-77521-5
DOI: 10.1007/978-3-642-77521-5

© Springer-Verlag Berlin Heidelberg 1993

The use of general descriptive names, registered names, trademarks, etc. in this publication does not imply, even in the absence of a specific statement, that such names are exempt from the relevant protective laws and regulations and therefore free for general use.

Product liability: The publishers cannot guarantee the accuracy of any information about dosage and application contained in this book. In every individual case the user must check such information by consulting the relevant literature.

Preface

Traditionally, in vitro diagnostics of cancer relies on the microscopic examination of tissue sections or cells. Numerous studies based on biochemical and immunochemical techniques focused on the metabolism, enzymes and structural components of tumor cells. Since qualitative differences between normal and tumor cells could not be detected, metabolic properties of human cells and surface-associated or shed tumor markers did not have a significant effect on the clinical differentiation between normal and malignant tissues.

Since the 1920s, a number of scientists have claimed that the cancerous cell is formed from normal cells by some kind of genetic mutation. For half a century, no methods were available to fully define the mutations on the molecular level. This changed dramatically, however, during the past 15 years. New techniques were developed and are continuing to be elaborated in the fields of molecular biology, molecular genetics, cytogenetics and cell biology. Application of some of these methods, such as transfer of human tumor DNA into cultured fibroblasts and hybridization, as well as DNA- and protein sequencing techniques, revealed homologies between retroviral oncogenes and normal human genes and paved the way to the first identification of specific genetic alterations in human tumors in 1982.

Since then many more genes altered in various malignancies have been identified. Starting from the observation that cellular genes homologous to retroviral oncogenes were located at breakpoints of chromosomal translocations, many genes at chromosomal breakpoints have been cloned which, because of homologies to structural motifs of proteins with known functions, e.g., transcription factors or homeobox genes, are thought to contribute to malignant transformation. Furthermore, the detection of chromosomal deletions by cytogenetic procedures or the use of polymorphic DNA markers led to the identification of genes which contribute to malignancy when genes or gene product are nonfunctional or lost.

A number of tumor-specific alterations in DNA structure, RNA synthesis and encoded proteins have been delineated. It can be anticipated that the confirmation of such alterations by enlarged studies on clinical specimens will gain significant diagnostic impact.

At present, established or putative areas of application have to be defined. Furthermore, a number of technical problems have to be solved before diagnostic procedures based on genetic alterations in cancer cells will become routine. In those cases in which alterations are of limited heterogeneity, the design of appropriate reagents and methods is straightforward. Some genes, however, harbor multiple mutations scattered over a significant portion of the gene. In these cases, rational approaches to the sensitive detection of a mutation in a given tumor are needed. Finally, the diagnostic significance of newly described cancer genes has not yet been established.

In order to delineate present and putative future applications of new diagnostic strategies based on the direct or indirect detection of mutated cancer genes, a conference on the "Application of Methods of Molecular and Cell Biology to the Diagnosis of Malignant Disease" was held in Hamburg, FRG, in October 1991, by the German Society for Clinical Chemistry. The conferene was sponsored by E. Merck, Darmstadt, FRG.

Methodological aspects such as enzymatic target amplification by the polymerase chain reaction, DNA fingerprinting and transfer of putative cancer genes to appropriate recipient cells were covered, in addition to recent aspects of the application of monoclonal antibodies in immunohistochemistry and immunoscintigraphy.

The characterization of oncogenes and tumor suppressor genes was correlated with growth control and dissemination of tumor cells on the basis of in vitro or clinical findings. The contributions in this book provide an update on established or newly described cancer genes and may help in the translation of concepts from basic research into clinical practice.

The ultimate goal of new diagnostic strategies in cancer is to improve diagnosis, prognosis and therapeutic regimens for patients. We sincerely hope that the present volume will bring us closer to that goal.

Hamburg/Darmstadt,   C. Wagener
Autumn 1992   S. Neumann

Contents

The Polymerase Chain Reaction:
Methodology and Diagnostic Potential (*A. Pingoud*) 1

Gene Loss as Marker in Tumor Development:
Considerations on the Origin of Specific Brain Tumors Due
to Combinant Loss of Function Mutations
of Tumor Suppressor Genes (*K. D. Zang*). 17

p53: Oncogene, Tumor Suppressor, or Both? (*W. Deppert*). . . . 27

Tumor Genome Screening by Multilocus DNA Fingerprints
as Obtained by Simple Repetitive Oligonucleotide Probes
(*J. T. Epplen, S. Bock, and P. Nürnberg*) 41

Activation of *ras* Oncogenes in Human Tumours
(*N. R. Lemoine*). 53

Laboratory Techniques in the Investigation
of Human Papillomavirus Infection (*E.-M. de Villiers*) 65

Detection of Minimal Residual Leukemia by Polymerase
Chain Reactions (*C. R. Bartram*) 77

A Dominant Metastogene Confers Metastatic Potential
to Tumor Cells (*H. Ponta, M. Zöller, K.-H. Heider, S. Seiter,
W. Rudy, M. Hofmann, C. Tölg, and P. Herrlich*). 87

Progression in Human Melanoma Is Accompanied
by the Altered Expression of Cell Adhesion Molecules
(*J. P. Johnson, B. G. Stade, U. Rothbächer, S. Stratil,
and G. Riethmüller*) 97

Structure, Function and Expression of the CEA Gene Family:
Diagnostic and Therapeutic Implications (*W. Zimmermann,
F. Grunert, G. Nagel, S. von Kleist, and J. Thompson*) 107

Membrane Proteins as Markers for Normal and Neoplastic
Endocrine Cells (*G. Lahr and M. Gratzl*) 117

Biological and Clinical Relevance of the Tumor-Associated
Serine Protease uPA (*M. Schmitt, F. Jänicke,
N. Chucholowski, L. Goretzki, N. Moniwa, E. Schüren,
O. Wilhelm, and H. Graeff*) 129

Analysis of Growth Regulatory Pathways in Human
Neuro-Oncology (*M. Westphal, W. Hamel, L. Anker,
and H.-D. Herrmann*) 151

HER-2/Neu/c-ERB-B2 Amplification: Clinical Relevance
in Mammary Carcinoma (*W. Jonat, K. Friedrichs,
H. Eidtmann, J. Meybohm, and S. Singh*) 169

Cytokine-Mediated Regulation of Growth Factor Receptors
(EGF-R and erb-B2) in Pancreatic Tumors (*H. Kalthoff,
C. Roeder, and W. Schmiegel*) 175

Chimeric CEA-Specific Antibody Fragments and Anti-idiotypic
Antibodies for Immunoscintigraphy and New Therapeutic
Approaches to Colorectal Carcinomas (*M. Neumaier*) 187

Tumor Localization by Immunoscintigraphy:
Potential and Limitations (*S. Matzku and H. Bihl*) 207

Subject Index 219

Contributors*

Anker, L. 151[1]
Bartram, C.R. 77
Bihl, H. 207
Bock, S. 41
Chucholowski, N. 129
Deppert, W. 27
Eidtmann, H. 169
Epplen, J.T. 41
Friedrichs, K. 169
Goretzki, L. 129
Graeff, H. 129
Gratzl, M. 117
Grunert, F. 107
Hamel, W. 151
Heider, K.-H. 87
Herrlich, P. 87
Herrmann, H.-D. 151
Hofmann, M. 87
Jänicke, F. 129
Johnson, J.P. 97
Jonat, W. 169
Kalthoff, H. 175
Kleist, S. von 107
Lahr, G. 117
Lemoine, N.R. 53
Matzku, S. 207

Meybohm, J. 169
Moniwa, N. 129
Nagel, G. 107
Neumaier, M. 187
Nürnberg, P. 41
Pingoud, A. 1
Ponta, H. 87
Riethmüller, G. 97
Roeder, C. 175
Rothbächer, U. 97
Rudy, W. 87
Schmiegel, W. 175
Schmitt, M. 129
Schüren, E. 129
Seiter, S. 87
Singh, S. 169
Stade, B.G. 97
Stratil, S. 97
Thompson, J. 107
Tölg, C. 87
Villiers, E.-M. de 65
Westphal, M. 151
Wilhelm, O. 129
Zang, K.D. 17
Zimmermann, W. 107
Zöller, M. 87

---

* The addresses of the authors are given on the first page of each contribution.
[1] Page on which contribution begins.

The Polymerase Chain Reaction: Methodology and Diagnostic Potential

A. Pingoud

Institut für Biophysikalische Chemie, Zentrum Biochemie, Medizinische Hochschule Hannover, Konstanty-Gutschow-Straße 8, W-3000 Hannover 61, FRG

Introduction

Since its discovery in 1983 (for an entertaining account of the discovery see [1]), the polymerase chain reaction (PCR) has revolutionized the life sciences. The remarkable success of this method is due to the fact that it is simple, versatile, and robust. Most importantly, there is increasing interest in the target of the PCR, namely DNA (or RNA), the carrier of the genetic information, whose precise composition one needs to know in order to understand the physiology and pathophysiology of living organisms.

The PCR involves the *in vitro* amplification of a specific DNA sequence, starting from as little as a single DNA molecule and ending up with a billion copies. The reaction is based on the precise annealing and faithful extension of two oligodeoxynucleotide primers that flank and thereby define the target region to be amplified in the double-stranded DNA template. A PCR cycle consists of denaturation of the double-stranded DNA, annealing of the primers and their extension using a DNA polymerase. With denaturation of the products a new cycle begins. Repeated cycles of denaturation, annealing and polymerization result in the exponential amplification of specific DNA fragments whose 5' ends are represented by the primers used in the polymerization. The length of the DNA fragments produced in the PCR is determined by the length of the primers used and the distance between the 3' ends of the primers when hybridized to the target DNA. Thus, product identification can make use of the length of the PCR products. Sequence information characteristic for the amplified DNA between the two primers, as well as "chemical tags" associated with the two primers and incorporated into the DNA fragments, can also be used for this purpose. The result of a PCR analysis can be a qualitative one, as needed for the detection of hereditary diseases and genetic polymorphisms (including disease predisposition), the identification of somatic mutations (including cancer) and the presence of viral or microbial pathogens. There is an

Wagener/Neumann (Eds.)
Molecular Diagnostics of Cancer
© Springer-Verlag Berlin Heidelberg 1993

increasing demand for quantitative PCR, for example to measure the "titer" of a pathogen, the number of infected or mutated cells in a tissue, and the concentration of certain mRNAs.

In the present contribution, I shall briefly describe the basic methodology and review recent developments regarding PCR procedures, from sample preparation to product identification. I shall also discuss improvements in PCR machines, as well as the results of efforts to automatize PCR analysis. I will briefly mention other amplification protocols as alternatives to PCR, such as the ligation chain reaction (LCR) and self-sustained sequence replication (3SR). Finally, some relevant recent applications of the PCR for medical diagnostic purposes will be mentioned.

It must be pointed out that two excellent review articles, similar in scope to this one, but with different emphasis, were published in 1991 [2, 3]. For a thorough introduction to the methodology and applications of the PCR, the reader is referred to four review volumes published in the last two years [4–7].

The Polymerase Chain Reaction

Template

The usual PCR template is double-stranded DNA. Single-stranded DNA or RNA can be used, the latter requiring reverse transcription. Typically, nanogram amounts of cloned DNA or microgram amounts of human genomic DNA are employed, the equivalent of 10^4–10^5 target DNA molecules. However, much lower amounts of DNA can be amplified, down to the DNA of single cells. In general, a standard nucleic acid purification involving a proteinase K treatment, phenol-chloroform extraction and ethanol precipitation is sufficient for most purposes. For special purposes, depending on the amount and complexity of the biological sample, more simple (lysing cells by boiling in water) or more complicated (separation of cells, removal of cell debris etc.) DNA extraction protocols have been worked out. For example, DNA preparation from blood previously were considered to be dependent on the separation of leukocytes from the vast excess of erythrocytes by a Ficoll-Hypaque density gradient centrifugation; now, protocols have been worked out in which erythrocytes are selectively lysed, then the DNA is extracted from the leukocytes by boiling in the presence of Chelex-100, which removes the iron ions [8]. Another example is given by the PCR amplification of DNA contained in paraffin-embedded fixed tissues for prospective studies. The success of the amplification depends on the fixative used, the time of fixation and the length of the target DNA. While formaldehyde fixation quickly leads to material that cannot be amplified anymore, alcohol fixation allows amplification after long storage times without affecting histological detail [9]. Other examples that demonstrate the robust nature of the PCR include amplification of DNA extracted from archival material, conserved ancient tissues [10] or autoclaved infectious organisms [11] and, most intriguingly of DNA inside cells [12].

It is noteworthy that the PCR (when carried out with the *Taq* polymerase) is "tolerant" of impurities in the DNA sample, in particular common organic solvents like ethanol, DMSO, DMF and formamide in concentrations up to 10%, urea in concentration up to $2M$ and many nonionic detergents in concentrations up to 5%; not, however, of ionic detergents like SDS [13].

Primers

The efficiency and the specificity of the PCR is dependent on the design of the primers. Most often the PCR works satisfactorily when the following requirements are fulfilled: Primers should be between 15 and 30 nucleotides in length (depending on the complexity of the DNA template and the T_M of the template primer hybrid), have an approximately 50% G + C composition and possess a similar T_M when hybridized to their target. They should not be complementary at their 3' ends (to avoid primer-dimer artifacts) and should not contain runs of C's or G's or self-complementary sequences.

Primers are always used in high excess over template, usually at a concentration of $0.1-1 \mu M$ each. For asymmetric PCR one primer is employed in approx. 100-fold excess over the other.

The versatility of the PCR is in part due to the fact that novel sequences can be generated and that "chemical tags" can be introduced by the PCR with special-purpose primers, like mismatch primers to produce site-directed mutants and primers that carry noncomplementary sequences at their 5' ends (restriction sites, phage RNA polymerase promotor sequences, "GC clamps", fluorescent reporter groups, a biotin moiety, a hapten etc.).

The Polymerase

The PCR was developed with the Klenow fragment of the *Escherichia coli* polymerase I, an enzyme that is irreversibly inactivated at elevated temperatures and, therefore, had to be added at every cycle after denaturation [14]. The thermolability of this enzyme, furthermore, precluded high annealing temperatures required for a highly specific template–primer interaction. The introduction of the thermostable DNA polymerase isolated from *Thermus aquaticus* had tremendously improved the usefulness of the PCR reaction for all purposes [15].

The *Taq* polymerase has a molecular weight of 94 kDa. It is stable at 93–95°C (half-life at 93°C approx. 120 min, at 95°C approx. 40 min) and, therefore, has to be added only at the beginning of the PCR. *Taq* polymerase has a temperature optimum of 75°C, which guarantees highly specific template–primer annealing and efficient extension even with templates prone to form secondary structures. Elongation rate is of the order of 100 nucleotides/s at 75°C. Lowering the temperature to 40°C reduces the elongation rate by two orders of magnitude. This implies that appreciable primer extension takes place even during annealing, which usually is carried out at temperatures between 40°C and 75°C. With 50–60 nucleotides incorporated per binding event the *Taq* polymerase is not as processive as originally believed. This means that incomplete, truncated products are produced which can be exploited for direct sequencing of PCR products [16] but which also necessitate electrophoretic purification of the desired product for some applications. The *Taq* polymerase has a 5′–3′ but no 3′–5′ exonuclease activity. The absence of a proofreading activity in *Taq* polymerase explains why this enzyme is less accurate than, for example, the Klenow enzyme or the T4 and T7 polymerase, which have a 3′–5′ exonuclease activity. The error rate of approximately 10^{-4} per nucleotide and cycle, as measured under standard conditions with various templates [17], will not interfere with most applications that utilize the population of amplified DNA molecules. Problems arise when the PCR is carried out with a small amount of target DNA, because then errors introduced in the first few rounds of replication are amplified. The infidelity of the *Taq* polymerase also must be taken into account when PCR products are cloned or, in general, when the focus is on individual DNA molecules present in the PCR product mixture, for example when a quantitative PCR is carried out to determine the mutant frequency in a population of cells. Efforts have been undertaken to optimize the conditions of the PCR with regard to fidelity. By systematic variation of the concentration of deoxynucleoside triphosphates and Mg^{2+}, as well as pH, conditions were found under which the *Taq* polymerase showed threefold higher accuracy, albeit at lower efficiency [18]. Recently, a variant of the *Taq* polymerase has been introduced: the Stoffel fragment, a N-terminal truncated 61-kDa protein which lacks 5′–3′ exonuclease activity, shows high temperature stability (Stoffel fragment half-life 20 min at 97.5°C, *Taq* polymerase half-life 10 min at 97.5°C) and displays a broader Mg^{2+} optimum than native *Taq* polymerase. It is recommended for amplification of targets rich in G + C content or containing complex secondary structures.

Other thermostable polymerases have been identified and isolated. Of particular interest, because of their commercial availability and their properties, are the Vent DNA polymerase isolated from *Thermococcus litoralis* and the *Pfu* DNA polymerase isolated from *Pyrococcus furiosus*. The Vent DNA polymerase is considerably more thermostable than the *Taq* polymerase: its half-life at 100°C is approx. 90 min. Presumably because it has an associated 3′–5′ exonuclease activity, the Vent polymerase is more accurate than the *Taq* polymerase: under optimized conditions it shows approx. twofold higher fidelity and amplification efficiency than the *Taq* polymerase [18]. The extremely thermostable *Pfu* polymerase has been advertised to be even more accurate, which makes it an interesting enzyme for cloning of PCR-amplified DNA and the analysis of genetic polymorphisms on a variable background.

Although *Taq* polymerase can transcribe RNA into DNA, this activity is too weak to be useful for a coupled reversed transcription (RT)/amplification reaction. This is different with the analogous (same size, high homology) enzyme from *Thermus thermophilus*, in the presence of $1\,mM$ $MnCl_2$ an efficient reverse transcriptase, which can cope with RNA templates with a high degree of secondary structure. By chelating Mn^{2+} by EGTA and supplying Mg^{2+} it can efficiently and accurately amplify the resulting cDNA [19]. It seems that the *Tth* polymerase will be very useful for the detection, quantitation and cloning of cellular and viral RNA.

Reaction Conditions and Cycling Parameters

Reaction conditions and cycling parameters should be optimized for each particular application. The following buffer has been found useful for most applications employing *Taq* polymerase: $10\,mM$ Tris·HCl pH 8.3 (at 25°C), $50\,mM$ KCl, $1.5\,mM$ $MgCl_2$, 0.01% (w/v) gelatin. For G + C-rich templates that are not easily denatured, or templates with pronounced secondary structures that hinder the extension of primers, it may help to include cosolvents like DMSO (1%–10%) or glycerol (10%–20%) into the amplification mixture [20]. The dNTP concentration should be in the order of $200\,\mu M$, i.e., well above the *Taq* polymerase K_M of $10-15\,\mu M$. dNTP and Mg^{2+} concentrations are most critical for the efficiency and accuracy of the PCR (*vide supra*).

The temperature vs time profile for the amplification process depends primarily on the T_M of the template–primer hybrid, the length of the DNA to be amplified and the thermocycler used (because the desired temperature can considerably deviate from the actual one). A standard PCR protocol with *Taq* polymerase is: 1 min at 90–95°C for denaturation, 1 min at 40–75°C for annealing and 2 min at 65–75°C for extension. Longer templates may need longer extension times, and templates with very stable secondary structure may require a 99°C denaturation step. The PCR can be carried out as a two-step process, with only denaturation and subsequent annealing/extension. The cycle time of the PCR, usually around 4 min, could be much shorter, if the temperature change in the block and the heat transfer between block, tube and solution could be made faster and more effective (*vide infra*). The number of cycles, of course, depends on the initial concentration of the target in the sample and the amount of product one wants to obtain. With an duplication efficiency of close to two per cycle a 10^6-fold amplification is achieved after 20 cycles and a 10^9-fold amplification after 30 cycles. Limiting for the PCR usually is the polymerase concentration. With excess dNTPs and primers amplification is exponential until the template concentration approaches the polymerase concentration. From then on amplification is linear. The "plateau" of the product vs time profile is characterized by the lack of availability of free polymerase, the competition of complementary strand and primer reannealing to the template, the inactivation of the polymerase, the accumulation of pyrophosphate and, finally, the consumption of primers and dNTPs.

The proper choice of the reaction conditions and cycling parameters determines the specificity of the PCR with a given template and primer pair. One reason for nonspecific product formation could be that during the initial heating step primers bound nonspecifically to single-stranded regions are already elongated and, thereby, stabilized such that they do not dissociate from the template during the elongation step. Other nonspecifically annealed and elongated primers on the complementary strand, when close to each other, could lead to nonspecific products, including "primer dimers". This problem can be overcome by adding the polymerase to the reaction mixture after the first annealing step. Another way of minimizing nonspecific amplification is to use two pairs of primers. The first pair (outer primers) defines a stretch of DNA which is amplified first, while the second pair (inner or nested primers) is complementary to a region within the DNA amplified first and leads to amplification of this piece of DNA. While the two rounds of amplification can be separated by simply diluting the first amplification mixture into the second one, a more elegant and safer approach is to take advantage of differential stabilities of GC-clamped outer primers and nested primers when hybridized to the template. Such a PCR protocol begins with a high annealing temperature to amplify the target of the outer primers; the annealing temperature and subsequently also the denaturation temperature is then dropped after several cycles to preferentially amplify the target of the inner primers [2].

It must be kept in mind that a lack of specificity in the amplification process can only be compensated by specific detection systems when a significant proporation of the product is the desired one.

Prevention of Contamination and Carry-Over

Contamination by exogenous DNA is a serious problem when the PCR is carried out with a few DNA templates. Some of these contaminations can be avoided by following good laboratory practice, like pre-aliquoting reagents, using positive displacement pipettes, and reserving separate areas for pre-PCR (sample preparation), PCR and post-PCR (product analysis) work. Nevertheless, single "replicons" can escape and may contaminate a sample, leading to false-positive signals. Pre-PCR sterilization of the reaction mixture (without enzyme and template) by photochemically modifying and thereby inactivating the contaminating DNA can be accomplished by short-wavelength UV irradiation [21]. Contamination of the target DNA and the enzyme, which is most serious with the products of previous PCRs in which similar material was amplified, cannot be prevented by this procedure. To cope with this problem, two very effective schemes have been developed. Post-PCR sterilization of amplified DNA can be achieved by carrying out the PCR in the presence of isopsoralens which, when irradiated with long-wavelength UV light, form covalent adducts with the amplified DNA. This modification abolishes the activity of the DNA to be used as a template but does not interfere with hybridization reactions [22]. Another post-PCR sterilization protocol makes use of the fact that dUTP can substitute for dTTP in the polymerization reaction and that DNA containing deoxyuridylic acid residues is attacked by uracil-*N*-glycosylase to produce abasic sites that lead to cleavage of the DNA upon heat treatment at alkaline pH. Consequently, when PCR reaction mixtures are treated with uracil-*N*-glycosylase prior to amplification, contaminating DNA from previous reactions is destroyed. As uracil-*N*-glycosylase is irreversibly denatured by elevated temperatures it does not interfere with the new PCR [23].

Thermocyclers

The first generation of microprocessor-controlled thermocyclers is currently being replaced. Heating and cooling in the new thermocyclers is done by Peltier elements. Low-mass sample blocks with low heat capacity and high heat conductivity are employed to achieve fast and accurate temperature changes with minimal overshoot. They use a microtitre plate format which can accommodate up to 8×12 tubes which are specially designed to optimally fit into the block, to eliminate the need for oil in the sample wells, and which have thin walls for fast heat transfer. Heated covers prevent condensation of water from the sample, allowing the PCR to be carried out without a mineral oil overlay.

While the second generation of thermocyclers promise to reduce amplification time by a factor of 3 to 4, a standard 25-cycle PCR will still require more than an hour. This is not the "biochemical" limit, because the cycle of denaturation, annealing and extension can be completed in less than 10 s with a 500-bp target. Indeed, it has been shown with a rapid air thermocycler and a 10-µl sample sealed into a thin glass capillary that a 30-cycle amplification of a 536-bp β-globin DNA fragment starting from 50 µg human genomic DNA can be carried out within 15 min [24].

Analysis of PCR Products

The qualitative and quantitative analysis of PCR products usually requires some confirmation that the product obtained is the desired one. This can be done by identifying the length and/or sequence of the product.

Qualitative Analysis Using Size Information

Product detection is most often accomplished by ethidium staining of PCR products separated by agarose or polyacrylamide gel electrophoresis. A typical PCR yields approx. 1–100 ng of product which is sufficient for ethidium staining. Smaller amounts can be detected by silver staining or, most sensitively, by Southern analysis using radioactively or otherwise labeled probes which, furthermore, allow under stringent hybridization conditions to differentiate between genetic variants due to point mutations. Hardly used so far for product detection and identification is HPLC on columns separating according to size and coupled to highly sensitive UV detectors; HPLC analysis is likely to be the method of choice in the near future, because it is fast and lends itself to automation and quantitation much better than an electrophoretic analysis.

The detection of genetic variations (mutations, polymorphisms) by the PCR usually is done with prior knowledge of what the variation is like. Several protocols have been developed for this purpose. Insertions and deletions are most easily detected, because they lead to a different size of the amplification product which can be analyzed by gel electrophoresis. When variable insertions and deletions occur in large genes multiplex amplification of several regions can be carried out simultaneously, as demonstrated for deletion detection in the Duchenne's muscular dystrophy locus [25]. This, however, necessitates testing many different primer pairs before all regions are amplified to a similar extent. Reciprocal translocations can be similarly identified on the DNA level when the breakpoints on both chromosomes occur within a small well-defined region [26] or on the mRNA level after reverse transcription using primers from the 5′ region of one chromosome and the 3′ region of the other [27].

The detection of point mutations requires more sophisticated techniques. The most straightforward method to detect point mutations, allele-specific PCR (ASPCR), relies on the low efficiency with which primers carrying a mismatch at their 3′ end can be elongated. Thereby, normal and mutant DNA can be differentiated using primers which are at their 3′ end complementary to one or the other. With the primer that is specific for the normal DNA significant product formation is only seen with the normal DNA, not with the mutant DNA, and *vice versa* with the primer that is specific for the mutant DNA. If in a DNA sample both the normal and the mutant templates are present, both primers will give rise to amplified products. The method, however, does not work satisfactorily with all mismatches, because some primers are elongated even when they are not perfectly matched at the 3′ end, which leads to ambiguous results.

Alternatively, primers can be designed such that they generate an artificial restriction site with the normal or the mutant DNA template. The amplified DNA is analyzed by electrophoresis after it has been digested with the restriction enzyme specific for the site introduced. The RFLP is indicative of the particular mutation and can even be used for quantitative purposes [28].

When the amplified DNA is hybridized with a complementary RNA or DNA probe the site of mutation can be detected by identifying the mismatch by enzymatic (RNase A) or chemical (hydroxylamine and osmium tetroxide) cleavage of the heteroduplex and analyzing the cleavage products on denaturing gels. The advantage of this method is that it does not require knowledge of the site and kind of mutation. However, it fails to detect all mutations reproducibly [29].

More successful in this respect is a technique called GC-clamped denaturing gradient gel electrophoresis (DGGE) [30]. It takes advantage of the fact that a single point mutation alters the stability of double-stranded DNA or DNA·RNA hybrids which can be detected by electrophoresis on polyacrylamide gels containing a linearly increasing gradient of formamide or urea. This technique is particularly sensitive when one end of the duplex is stabilized by 20–40 GC base pairs introduced by one of the primers. This GC clamp is the most stable region of the molecule and, therefore, allows detection of partial denaturation of the duplex in the region of the mutation. DGGE also does not require knowledge of the site of mutation. A variant of this method employs a temperature gradient instead of the chemical denaturant gradient; this temperature gradient gel electrophoresis (TGGE) has been employed with modifications for quantitative PCR [31].

Sequence-dependent variations in the electrophoretic mobility of single strands can be picked up by high-resolution polyacrylamide gel electrophoresis under nondenaturing conditions [32]. Whether the single-strand conformation polymorphism (SSCP) can resolve as many genetic variants as the DGGE or TGGE technique remains to be established.

Qualitative Analysis Using Sequence Information

By far the most successful protocols to analyze and identify PCR products employ oligodeoxynucleotide probes for sequence-dependent detection. These protocols, in one variation or the other, involve annealing of a labeled oligodeoxynucleotide probe to a target sequence located somewhere between the sequences complementary to the PCR primers. Dot-blot analysis of immobilized amplified DNA using radioactively labeled allele-specific oligodeoxynucleotides (ASO) were first used for the identification of normal and mutant β-globin gene sequences [33, 34]. In the reverse dot-blot format, the ASO probe is immobilized and hybridized with the amplified DNA [35]. When the amplified DNA carries a chemical tag, e.g., a biotin moiety, it can be detected, bound to the immobilized ASO probe, by a color reaction, e.g., using a streptavidin–horseradish peroxidase conjugate and a chromogenic substrate. The reverse dot-blot procedure is particularly effective when several ASO probes specific for various mutations are immobilized on one strip, which allows the typing of a genetic disease with several variants, like β-thalassemia, or identification of genetic poly-morphisms at particular loci, e.g., the HLA DQα locus.

Sandwich capture techniques employing biotinylated oligodeoxynucleotide have been used to develop an automated system to detect point mutations [36]. This system, which demonstrates the state of the art in DNA diagnostics, employs two directly adjacent oligonucleotides complementary to a sequence within the PCR-amplified DNA; one is a 5′ biotinylated probe with its 3′ end at the site of mutation, the other is a probe carrying a suitable reporter group at the 3′ end, e.g., a digoxigenin residue. The two oligodeoxy-nucleotides are hybridized to the target DNA and ligated, provided they are perfectly base paired. Capture of the biotinylated ligation product on immobilized streptavidin and analysis for the 3′ reporter group allows detec-tion of point mutations without electrophoresis or centrifugation [36].

PCR and DNA Sequencing

The PCR is often carried out with the purpose of sequencing the amplified product, e.g., in order to identify a mutation. PCR products can be sequenced by subcloning into standard sequencing vectors and then follow-ing an established dideoxy sequencing protocol. The problem with this procedure is that during the PCR errors accumulate, such that several cloned PCR products have to be sequenced in order to obtain a consensus which reflects the mutation in the original genomic sequence and not the random mutations introduced by the polymerase during the amplification process. Sequencing the entire population of PCR products directly, either by the dideoxy method or by chemical means, circumvents the problem associated with subcloning the amplified DNA and sequencing isolated clones. A variety of protocols have been developed for this purpose (for a review see [37]). Most of them depend on the production of single strands. This can be achieved by asymmetric PCR with one primer in high excess over the other. Alternatively, a biotinylated primer can be used in the PCR which allows affinity purification of one strand on an avidin or streptavidin matrix.

Quantitative PCR

There is an increasing demand for a quantitative PCR to determine virus or parasite titers, to measure the number of transformed cells in a given tissue and to monitor gene expression, i.e., to determine the concentration of mRNAs of interest etc. In principle, during the exponential phase of amplification there is a linear relationship between the initial concentrations of target present in various samples and the concentrations of product obtained with these samples during the PCR, provided the reaction is carried out under exactly the same conditions. On approaching the plateau phase, the linear relationship no longer holds. Nevertheless, there is still a proportionality between initial target concentrations and product concentrations [38, 39]. While very pure DNA samples can be quantitated in the exponential phase and the preplateau phase, this cannot be achieved in a reproducible manner with typical biological samples, due to heterogeneity in composition which may influence the annealing or polymerization kinetics. Internal standardization, therefore, is mandatory for crude DNA samples. Coamplifying an internal standard has the additional advantage that quantitation can be carried out even in the plateau phase, where the signal is much stronger. Internal standards should be as similar as possible to the target to be analyzed. Ideal is a mutant version of the target of interest, for example a DNA carrying a point mutation or a short deletion or insertion. The target to be analyzed is coamplified with the internal standard, which is added to the PCR mixture before the amplification is started. Coamplification is best carried out with the same primer pair. If, however, the detection system requires two different primer pairs, again these should be as similar as possible.

Simultaneous detection of amplified target and internal standard can be incorporated into size- and/or sequence-dependent detection systems. For example, PCR can be combined with TGGE to quantitate low copy numbers of DNA (or RNA, after reverse transcription) using an internal standard which differs from the target by a single internal point mutation [31]. After coamplification a small amount of labeled standard is added. After denaturation and renaturation the labeled standard has formed homoduplexes with the amplified standard and heteroduplexes with the amplified target, in a ratio reflecting the relative concentration of target and added standard before amplification. Homoduplexes and heteroduplexes are separated by TGGE and quantitated. The method has high sensitivity, excellent precision and a broad dynamic range. Another example is a sequence-dependent format to detect simultaneously amplified target and internal standard [40]. In this procedure a biotinylated primer is used to capture the amplified DNA on a streptavidin-coated matrix. The internal standard, which is added to the reaction mixture in serial dilutions prior to the amplification, differs from the target by having an insert comprising the *lac* operator sequence. After the PCR, which in this study was carried out with nested primers for high specificity, the amplified target and standard DNA is bound to the streptavidin-coated matrix. The amount of amplified standard and, thereby, indirectly the relative amount of target is quantitated by a colorimetric assay employing a β-galactosidase LacI repressor fusion protein and a chromogenic substrate. A great advantage of this method is that it can be easily automated.

Alternatives to the PCR

The concept of *in vitro* gene amplification by enzymatic means, once introduced and shown to work with DNA polymerases, prompted scientists to look for alternatives (if for no other reason than to circumvent the patent problems associated with the PCR). The most successful alternative so far is the ligase chain reaction (LCR) (for a recent review see [41]). In the LCR two oligodeoxynucleotide primer pairs complementary to both strands of the target DNA are annealed and joined by DNA ligase. After denaturation the ligation product serves as a template in the next cycle, such that, as with the PCR, exponential amplification results. As a thermostable ligase from *Thermus aquaticus* has been discovered, the LCR can be carried out in a thermocycler with a single addition of *Taq* ligase at the beginning of the amplification reaction. The LCR is not as versatile as the PCR, because only small sequences can be amplified. This limits the usefulness of this reaction to the detection of point mutations. It can, however, also be used in combination with the PCR, which may extend its range of applications.

Self-sustained sequence replication (3SR) also is a nucleic acid sequence-based amplification technique which consists of continuous cycles of reverse transcription and transcription using a double-stranded cDNA as an intermediate (for a recent review see [42]). When an RNA target is to be amplified it is first transcribed by reverse transcriptase using a primer A that introduces a T7 RNA polymerase promotor. After RNase H treatment and formation of the double-stranded cDNA intermediate by reverse transcriptase using a primer B, the target sequence is transcribed many times from the cDNA by T7 RNA polymerase to give an RNA complementary to the initial target RNA. The antisense RNA is immediately converted to cDNA, thereby starting a new cycle. The 3SR protocol can be adapted to amplify DNA target. Different from the PCR and the LCR, transcription-based amplification systems require three enzyme activities and work in a continuous fashion at a constant temperature (with AMV reverse transcriptase, *Escherichia coli* RNase H and T7 RNA polymerase at 42°C). At present, the 3SR technique is less specific than PCR and LCR because the enzymes employed are not thermostable, which does not allow highly stringent hybridization conditions. On the other hand it is much faster than PCR and LCR and does not require thermal cycling.

Diagnostic Applications of the PCR

The impact of the PCR technique on the diagnosis of human genetic diseases has been enormous (for a review see [43]). It has to a large extent replaced Southern analysis and, because of its sensitivity, has opened new areas for clinical diagnosis. With its many variants that allow for easy automation it is going to be a routine procedure in the clinical laboratory for the identification of the more common hereditary diseases. Laboratory kits are already being developed for the diagnosis of several hemoglobinopathies, bleeding disorders and cystic fibrosis.

In addition to the diagnosis of genetic diseases, the PCR has facilitated the analysis of hereditary disease susceptibility, for example those which are linked to certain HLA haplotypes, like type I diabetes and certain types of rheumatoid arthritis.

A rapidly expanding application of the PCR concerns the identification of mutations of oncogenes and anti-oncogenes. PCR analysis of the *bcr/abl* fusion is increasingly used for the diagnosis of chronic myelogenous leukemia and, after treatment, for the detection of residual affected cells [27, 44]. Similarly, PCR is employed for the detection of *ras* point mutations in small subpopulations of cells for an identification of tumor progression markers [28, 45]. Likewise, the reduction to homozygosity involving the *p53* anti-oncogene has been demonstrated by the PCR to be involved in certain types of cancers [46]. It can be envisaged that testing tumor tissues for activated oncogenes and inactivated anti-oncogenes by PCR methods will greatly expand our knowledge of tumorigenesis.

The detection of specific viral, bacterial and other pathogens associated with various infectious diseases, so far mainly carried out by immunological techniques, will in part be taken over by PCR methods. This is, for example, the case with human immune deficiency virus (HIV) detection, particularly in newly infected patients who have not yet seroconverted, for which a quantitative PCR method was developed [47]. For some other pathogens no convenient, rapid detection methods exist. This is, for example, the case with human papilloma virus (HPV) infections, which are among the most common sexually transmitted viral diseases and have been associated with many human cancers. Here, PCR methods allow detection of the HPV infection and efficient differentiation between various subtypes [48, 49]. PCR methods are now routinely also employed to identify cytomegalovirus (CMV), hepatitis B virus (HBV) and certain enteroviruses.

Finally, a major application of the PCR for medical purposes concerns tissue typing for organ and tissue transplantation. Kits for the analysis of the HLA DQα locus are already commercially available. Related to this application is the use of PCR for forensic purposes. It is especially in this area that the PCR can exploit its extreme sensitivity.

Acknowledgments. PCR work in the author's laboratory was carried out with the financial support of E. Merck, Darmstadt, the Deutsche Forschungsgemeinschaft (Pi122/5-1), and the Fonds der Chemischen Industrie. Stimulating discussions with many colleagues, in particular Dr. Axel Landgraf, are gratefully acknowledged. Thanks are due to Mrs. E. Schuchardt for typing the manuscript.

References

1. Mullis KB (1990) The unusual origin of the polymerase chain reaction. Sci Am 262:36–44
2. Erlich HA, Gelfand D, Sninsky JJ (1991) Recent advances in the polymerase chain reaction. Science 252:1643–1651
3. Bloch W (1991) A biochemical perspective of the polymerase chain reaction. Biochemistry 30:735–744
4. Erlich HA (ed) (1989) PCR technology. Stockton, New York
5. Erlich HA, Gibbs R, Kazazian HH Jr (eds) (1989) Polymerase Chain Reaction. Cold Spring Harbor, New York
6. Innis MA, Gelfand DH, Sninsky JJ, White TJ (1989) PCR protocols. Academic, San Diego
7. Marx P (ed) (1990) PCR topics: usage of polymerase chain reaction in genetic and infectious diseases. Springer, Berlin Heidelberg New York
8. Winberg G (1991) A rapid method for preparing DNA from blood suited for PCR screening of transgenic mice. PCR Meth Appl 1:72–74
9. Greer CD, Lund JK, Manos MM (1991) PCR amplification from paraffin embedded tissues: recommendations on fixatives for long-term storage and prospective studies. PCR Meth Appl 1:46–50
10. Pääbo S (1989) Ancient DNA: extraction, characterization, molecular cloning, and enzymatic amplification. Proc Natl Acad Sci USA 86:1989–1943
11. Barry T, Gannon F (1991) Direct genomic DNA amplification from autoclaved infectious microorganisms using PCR technology. PCR Meth Appl 1:75
12. Haase AT, Retzel EF, Staskus KA (1990) Amplification and detection of lentivirus DNA inside cells. Proc Natl Acad Sci USA 87:4971–4975
13. Weyant RS, Edmonds P, Swaminathan B (1990) Effect of ionic and nonionic detergents on the Taq polymerase. BioTechniques 9:308–309
14. Saiki RK, Scharf S, Faloona F, Mullis KB, Horn GT, Erlich HA, Arnheim N (1985) Enzymatic amplification of β-globin genomic sequences and restriction sites analysis for analysis of sickle cell anemia. Science 230:1350–1354
15. Saiki RK, Gelfand DH, Stoffel S, Scharf SJ, Higuchi R, Horn GT, Mullis KB, Erlich HA (1988) Primer directed enzymatic amplification of DNA with a thermostable DNA polymerase. Science 239:487–491
16. Olsen DB, Eckstein F (1989) Incomplete primer extension during in vitro DNA amplification catalyzed by Taq polymerase; exploitation for DNA sequencing. Nucleic Acids Res 17:9616–9620
17. Eckert KA, Kunkel TA (1991) DNA polymerase fidelity and the polymerase chain reaction. PCR Meth Appl 1:17–24
18. Ling LL, Koehavong P, Dias C, Thilly WG (1991) Optimization of the polymerase chain reaction with regard to fidelity: modified T7, Taq and vent polymerases. PCR Meth Appl 1:63–69
19. Myers TW, Gelfand DH (1991) Reverse transcription and DNA amplification by a Thermus thermophilus DNA polymerase. Biochemistry 30:7661–7666
20. Smith KT, Long CM, Bowman B, Manos MM (1990) Using cosolvents to enhance PCR amplification. Amplifications:16–17
21. Sarkar G, Sommer SS (1990) Shedding light on PCR contamination. Nature 343:27
22. Isaacs ST, Tessman JW, Metchette KC, Hearst JE, Cimino GD (1991) Post-PCR sterilization: development and application to an HIV-I diagnostic assay. Nucleic Acids Res 19:109–116

23. Longo MC, Berninger MS, Hartley JL (1990) Use of uracil DNA glycosylase to control carry-over contamination in polymerase chain reactions. Gene 93: 125–128
24. Wittwer CT, Garling DJ (1991) Rapid cycle DNA amplification: time and temperature optimization. BioTechniques 10:76–83
25. Chamberlain JS, Gibbs RA, Ranier JE, Nguyen PN, Caskey CT (1989) Deletion screening of the Duchenne muscular dystrophy locus via multiplex DNA amplification. Nucleic Acid Res 16:11141–11149
26. Crescenzi M, Seto M, Herzige GP, Weiss PO, Griffith RC, Korsmeyer SJ (1988) Thermostable DNA polymerase chain amplification of t(14;18) chromosomal breakpoints and detection of minimal residue disease. Proc Natl Acad Sci USA 85:4869–4875
27. Kawasaki ES, Clark SS, Coyne MY, Smith SD, Chaplin R, Witte ON. McCormick FP (1988) Diagnosis of chronic myelogenous leukemia and acute leukemia by detection of leukemia-specific mRNA sequences amplified in vitro. Proc Natl Acad Sci USA 85:5689–5695
28. Haliassos A, Chomel JC, Grandjouan S, Kruh J, Kaplan JC, Kitzis A (1989) Detection of minority point mutations by modified PCR technique: a new approach for a sensitive diagnosis of tumor-progression markers. Nucleic Acids Res 17:8093–8099
29. Theophilus BDM, Latham T, Grabowski GA, Smith FI (1989) Comparison of RNaseA, a chemical cleavage and GC clamped denaturing gradient gel electrophoresis for the detection of mutations in exon 9 of the human β-glucosidase gen. Nucleic Acids Res 17:7707–7722
30. Sheffied VC, Cox DR, Lerman LS, Myers RM (1989) Attachment of a GC clamp to genomic DNA fragment by the polymerase chain reaction results in improved detection of single base changes. Proc Natl Acad Sci USA 86: 232–237
31. Henco K, Heibey M (1990) Quantitative PCR: the determination of template copy numbers by temperature gradient gel electrophoresis. Nucleic Acids Res. 18:6733–6734
32. Orita M, Iwahana H, Kanazawa H, Hayashi K, Sekiya T (1989) Detection of polymorphisms of human DNA by gel electrophoresis as single strand conformation polymorphisms. Proc Natl Acad Sci USA 86:2766–2770
33. Saiki RK, Bugawan TL, Horn GT, Mullis KB, Erlich HA (1987) Analysis of enzymatically amplified β-globin and HLA-DQα DNA with allele-specific oligonucleotide probes. Nature 324:163–167
34. Kazazian HH Jr, Boehm CD (1988) Molecular basis and prenatal diagnosis of β-thalassemia. Blood 72:1107–1116
35. Saiki KR, Walsh PS, Levenson CH, Erlich HA (1989) Genetic analysis of amplified DNA with immobilized sequences specific oligonucleotide probes. Proc Natl Acad Sci USA 86:6230–6234
36. Nickerson DA, Kaiser R, Lappin S, Stewart J, Hood L, Landegren U (1990) Automated DNA diagnostics using an ELISA-based oligonucleotide ligation assay. Proc Natl Acad Sci USA 87:8923–8927
37. Gyllensten UB (1989) PCR and DNA Sequencing. BioTechniques 7:700–705
38. Landgraf A, Reckmann B, Pingoud A (1991) Quantitative analysis of polymerase chain reaction (PCR) products using primers labeled with biotin and a fluorescent dye. Anal Biochem 193:231–235
39. Landgraf A, Reckmann B, Pingoud A (1991) Direct analysis of polymerase chain reaction products using enzyme linked immunosorbent assay techniques. Anal Biochem 198:86–91

40. Lundeberg J, Wahlberg J, Uhlén M (1991) Rapid colorimetric quantification of PCR amplified DNA. BioTechniques 10:68–73
41. Barany F (1991) The ligase chain reaction in a PCR world. PCR Meth Appl 1:5–16
42. Fahy E, Kwoh DY, Gingeras TR (1991) Self-sustained sequence replication (3SR): an isothermal transcription based amplification system alternative to PCR. PCR Meth Appl 1:25–33
43. Reiss J, Cooper DN (1990) Application of the polymerase chain reaction to the diagnosis of human genetic disease. Hum Genet 85:1–8
44. Martiat P, Maisin D, Philippe M, Ferrant A, Michaux JL, Cassiman JJ, van den Berghe H, Sokal G (1990) Detection of residual BCR/ABL transcripts in chronic myeloid leukaemia patients in complete remission using the polymerase chain reaction and nested primers. Br J Haematol 75:355–358
45. Chen J, Viola MV (1991) A method to detect ras point mutations in small subpopulations of cells. Anal Biochem 195:51–56
46. Meltzer SJ, Yin J, Huang Y, McDaniel TK, Newkirk C, Iseri O, Vogelstein B, Resau JH (1991) Reduction to homozygosity involving p53 in esophageal cancers demonstrated by the polymerase chain reaction. Proc Natl Acad Sci USA 88:4976–4980
47. Kellogg DE, Sninsky JJ, Kwok S (1990) Quantitation of HIV-1 proviral DNA relative to cellular DNA by the polymerase chain reaction. Anal Biochem 189:202–208
48. Sarkar FH, Crissman JD (1990) Detection of human papilloma virus DNA sequences by polymerase chain reaction. BioTechniques 9:180–185
49. Rodu B, Christian C, Snyder RC, Ray R, Miller DM (1991) Simplified PCR-based detection and typing strategy for human papilloma viruses utilizing a single oligonucleotide primer set. BioTechniques 10:632–636

Gene Loss as Marker in Tumor Development: Considerations of the Origin of Specific Brain Tumors Due to Combinant Loss of Function Mutations of Tumor Suppressor Genes

K.D. Zang

Institut für Humangenetik, Universität des Saarlandes, W-6650 Homburg/Saar, FRG

Results from the past decade of molecular genetics and cell biology research have confirmed the assumption that the loss of control over proliferation and differentiation of cells in fact is a multistep cascade of events. The models of tumorigenesis established by several investigators during the late 1960s and early 1970s have been experimentally verified in the meantime and are now, with a few modifications, widely accepted as the "Knudson model".

Based on his data on the epidemiology of retinoblastoma, an uncommon eye tumor of children, Knudson was able to demonstrate that the incidence of sporadic unilateral tumors exactly correlates with a two-hit mechanism and that of the bilateral and familial tumors with a one-hit mechanism. Thus, in the latter cases, one hit must be a germ-cell or germ-line mutation. In this context, it is remarkable that in 1971 Knudson began with the assumption of an additive impact of two nonallelic mutations which both cause a "dominant" step towards proliferation. This model of two or more gain-of-function mutations strikingly resembles today's data on activation of proto-oncogenes to oncogenes [6]. Also in 1971, Ohno was the first to present the idea of a recessive two-step mechanism by which, as he postulated, a proliferation control gene is lost first heterozygously, then homozygously [13]. His concept was correct but, sadly, he studied the wrong object. His ideas were based on the plausible, but now corrected assumption that the Philadelphia chromosome in chronic myelogenous leukemia was the result of a deletion by which a regulatory gene on the long arm of a G-group chromosome was lost. In fact, the Philadelphia chromosome is formed not by a deletion but by a reciprocal translocation between chromosome 22 and, in most cases, chromosome 9.

In 1973, Comings presented a "general theory of carcinogenesis" and established a very sophisticated two-step model of cancer development which is still valid today [1]. It is also based on sequential homozygous loss-of-function mutations of both alleles of a proliferation-controlling gene.

Wagener/Neumann (Eds.)
Molecular Diagnostics of Cancer
© Springer-Verlag Berlin Heidelberg 1993

Using the terminology of his time, Comings anticipated today's view of the working mechanisms of tumor suppressor genes. In a paper published in the same year, Knudson modified his original hypothesis in a similar way [7]. During the following years, he illustrated the validity of his model with a wide range of examples from clinical oncology.

It was back in 1967 that our group was the first to describe a uniform chromosome anomaly in a solid tumor. In meningioma, a benign tumor of the brain coverings, the complete loss of one G chromosome was shown to be a common event [21]. Later we identified the lost chromosome as chromosome 22 [22]. At that time we suggested a gene dosage effect caused by heterozygous loss of a proliferation control gene as the critical mechanism of the development of this slowly growing tumor.

Now we know that basically all authors were right. In virtually every kind of tumor we observe both gain of positive, i.e., proliferation-inducing signals and loss of negative, i.e., proliferation-suppressing signals. The cooperation and interaction of these signals determine the steps of tumor development from initiation through promotion and progression to metastasis. As a rule, considerably more than just two steps are needed [14]. One of the most detailed multistep models of tumorigenesis is Vogelstein's hypothesis of the development of colon carcinoma [4].

However, we learned that the "retinoblastoma paradigm" in its classical form, with the homozygous loss or inactivation of a down-regulating gene (the RB gene), can only be applied to retinoblastoma itself. Other tumors like Wilms' tumor and other candidate tumor suppressor genes like WT1 and p53 do not readily fit into this pattern, so that additional and modifying hypotheses are required (for reviews see [11, 19]).

Data from different fields of tumor research have provided evidence that there are indeed regulatory genes of which the physiological function is the suppression of cell proliferation or, under certain conditions, the induction or maintenance of differentiation:

1. Cell fusion experiments between tumor cells and normal cells showed that alleles which are lost in tumor cells can be replaced by normal alleles in a way that the tumor cells lose their malignant phenotype. In most instances, however, only a few parameters, such as growth rate or tumorigenicity in nude mice, can be reverted. A complete reversion of tumor cells to normal cells cannot be achieved by such means (for reviews see [5, 11]).

2. Findings from different chromosomes in many types of tumors under the heading "loss of heterozygosity" (LOH) demonstrate reproducible losses of RFLPs in the comparison of DNA samples from tumor vs constitutional DNA from the same patient. These observations, too, have to be treated with caution: they supply proof only of some kind of allelic loss; they provide no information if the lost allele is in fact the second one of the same gene, thus resulting in homozygosity, or the first mutation,

leading to a heterozygous state. This conclusion that LOH reveals a homozygous gene loss can be misleading because the consecutive mutations may well have happened at different loci (for review see [15]).

3. The observation of familial tumors with dominant transmission within an affected family, such as retinoblastoma or neurofibromatosis 1 and 2 (NF1, NF2), can be explained most conclusively by an inherited allele loss and a subsequent somatic mutation of the second allele of the same regulatory gene. This assumption, however, cannot explain the considerable inter- and intrafamilial heterogeneity of, for example, NF1, and, in hereditary cases of Wilms' tumor, it is incompatible with the molecular findings. The majority of (malignant) tumors require more than two mutations for their development, a fact which is not fully consistent with their apparently dominant pattern of inheritance.

Thus, several amendments have been added to the "retinoblastoma model" with respect to familial tumors like adenomatous polyposis, von Recklinghausen's disease (NF1) and other tumors which are at the focus of scientific progress. Recent cytogenetic and molecular data speak in favor of the involvement of several tumor suppressor genes, especially their combined heterozygous loss of function, rather than homozygous loss of only one tumor suppressor gene. With the exception of retinoblastoma, the epidemiology of which can be comprehensively explained by the homozygous loss of function of the RB gene, the remaining examples of recessive tumorigenic mutations are not convincing so far: Variability of symptoms and specific molecular features may point to genetic heterogeneity or to combined heterozygous loss of function of several suppressor genes, but also to interaction of tumor suppressor genes with viral genes at the product level. Interaction of a suppressor protein and an exogenous viral protein can cause a variable negative dosage effect. Genomic imprinting may mimic a heterozygous loss-of-function mutation. Interaction between normal and heterozygously altered suppressor proteins can give rise to a dominantly negative effect.

In the following, I would like to introduce a novel model of formal tumorigenesis which is based on the study of a complex of neuroectodermal tumors, namely meningioma, sporadic acoustic neuroma and familial neuroma (NF2), which are mostly benign tumors of the brain and spinal cord and the nerve sheaths respectively. In accordance with the epidemiologic, cytogenetic and molecular genetic knowledge about these tumors, the model proceeds from the existence and cooperation of two tumor suppressor genes which are both located on the long arm of chromosome 22. Furthermore, the model postulates a close genetic relationship between meningioma and acoustic neuroma which explains their frequently concomitant appearance in patients with NF2.

Instead of a rigid recessive mechanism, this model assumes a gene dosage effect which was discussed by us in a simpler form as early as 1967 and which explains in this context the expression of SV40 T antigen in a high percentage of meningiomas, as reported by us in 1975 and confirmed by several other groups [8, 20]. When our group described, for the first time in a solid tumor, a uniform chromosome aberration found in the majority of cases – namely the loss of a small acrocentric chromosome – neither oncogenes nor tumor suppressor genes were known at all. As mentioned, we argued then that a crucial gene for proliferation control must be lost with this chromosome, which we later identified as chromosome 22. Since meningioma is a very benign tumor, we presumed that the heterozygous loss of this gene suffices alone to give rise to slow but unlimited cell proliferation which, as a rule, goes on for years without further tumor progression. Such an assumption was supported by the fact that meningiomas are sporadic, nonfamilial tumors throughout, as well as by the uniform histologic appearance which resembles the proliferation centers of fetal brain coverings. In 1986, Seizinger et al. showed that in some cases of both meningioma and acoustic neuroma, RFLP analysis detects loss of heterozygosity of all genes on chromosome 22 [17, 18]. For a cytogeneticist, in meningioma this molecular proof of the well-known monosomy 22 or deletion of the long arm of this chromosome was trivial; no wonder these data were confirmed by all other groups, including ourselves [2, 12]. As to neuroma, this finding was new and interesting. Under the influence of Knudson's retinoblastoma paradigm, however, virtually all authors (over) interpreted this loss of alleles in meningioma and neuroma as proof of the homozygous loss of a tumor suppressor gene. However, unlike in retinoblastoma, no constitutional chromosome deletion has been observed. One interesting exception in meningioma has been difficult to interpret hitherto [10].

Although more and more polymorphic DNA probes have become available for chromosome 22, it has not been possible so far, by comparing parents and patient, to detect any constitutional deletion on the molecular level in any meningioma patient. Neither has it been possible to recognize a homozygous deletion of certain chromosome 22 sequences within the tumor tissue, nor to detect any allele loss in the subgroup of diploid meningiomas. Thus, not a single case is known so far in which a loss of heterozygosity on chromosome 22 gives convincing evidence of a retinoblastoma-like situation in which indeed a homozygous, not just a heterozygous gene loss has occurred by means of deletion of a defined chromatin structure on both copies of chromosome 22.

Furthermore, the allele losses published so far in meningioma, remarkably, have detected no interstitial deletion. Rather, they correspond to the known loss of a complete chromosome 22 or else the complete deletion of the distal two thirds of the long arm of chromosome 22. In neuroma and NF2, similar data were presented between 1986 and 1990. In the past 2 years, there have been surprising reports of the detection of interstitial deletions within the region proximal to the myoglobin locus in neuroma [16].

In the past few years, several groups have carried out molecular analysis of breakpoints in meningiomas and neuromas with only partial deletion of the long arm of chromosome 22. Their results, and even more the detailed evaluation of recent data on interstitial deletions of the long arm of chromosome 22, have led to the conclusion that in fact there are two different breakpoints, namely a proximal neuroma breakpoint and a more distal meningioma breakpoint, which apparently are separated by the myoglobin locus. This finding has led to the presumption that two tumor suppressor genes are located on the long arm of chromosome 22: an NF2 gene on the proximal side of the myoglobin locus and a meningioma gene located distally in the vast region between the myoglobin locus and the telomere [3, 16].

Neither the exact location nor the structure or function of these genes are known as yet. In any case, our initial hypothesis of a gene dosage effect, i.e., the assumption that the heterozygous loss of a negatively controlling gene, possibly in combination with other genetic defects, suffices alone to lead to a phenotypic effect, has not yet been falsified. Several authors now plead for the idea of such a dosage effect with respect to different types of tumors, e.g., Fearon and Vogelstein in their model of colorectal carcinogenesis regarding the p53 gene [4]. In neurofibromatosis 1, too, the combined occurrence of benign and malignant tumors and the considerable intrafamilial variability is now suggested to be the result of mutations of different tumor suppressor genes [9].

Remarkably, neuroma, unlike meningioma, occurs in a sporadic as well as in a familial form. The latter, also known as the dominantly heritable NF2, forms bilateral acoustic neuromas which frequently are accompanied by meningiomas.

It is remarkable that the loss of a long arm or a complete chromosome 22, which evidently plays the central role in meningioma and neuroma, is a frequently observed phenomenon in the course of progression of other neuroectodermal tumors and, moreover, is involved in a number of non-neuroectodermal tumors.

The assumption of two neighboring tumor suppressor genes on the long arm of chromosome 22, based on molecular genetic findings along with clinical and epidemiologic features of meningioma and neuroma, led me to draw up a speculative model of combined action of both genes (Figs. 1–3). Evidently this model is suitable to explain the epidemiologic findings:

- Meningioma and neuroma are, as a rule, benign and sporadic tumors occurring in older people.
- Meningiomas are considerably more frequent than neuromas.
- In a small percentage of cases, meningiomas occur as multiple tumors, but (with the possible exception of a very small number of cases) not as familial tumors.
- Neuromas, on the other hand, occur quite frequently as multiple tumors

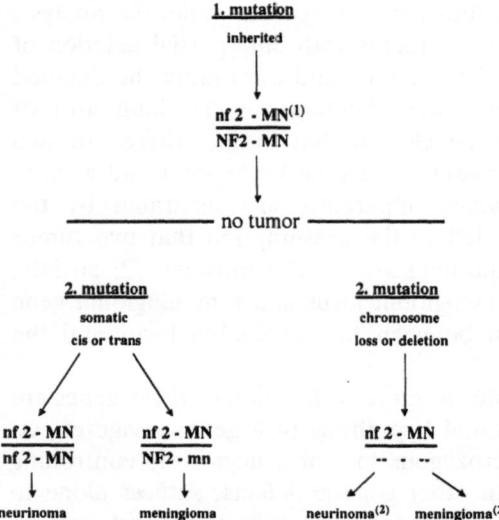

Fig. 1. Hypothetical molecular origin of neurinomas and meningiomas in NF2 kindreds. (*1*) In contrast to the NF2 gene, deletion or loss-of-function mutation of the MN gene may be incompatible with differentiation or survival of germ cells (almost no meningioma families). (*2*) Homozygous deletion or loss of the NF2 gene plus heterozygous loss of the MN gene may modify the type and/or grade of the tumor. Homozygous loss of the MN gene may be incompatible with survival or proliferation of somatic cells

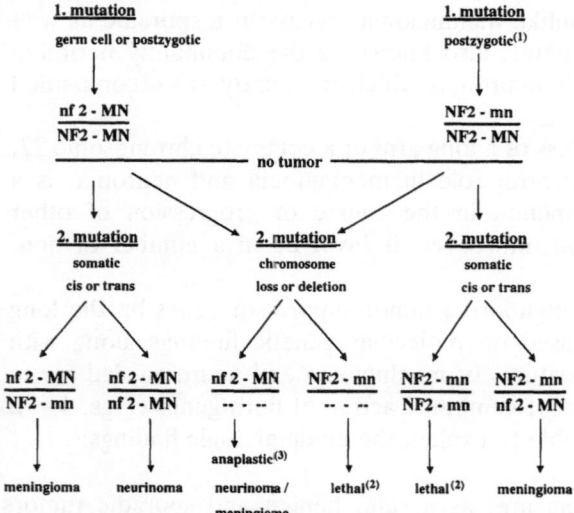

Fig. 2. Hypothetical molecular origin of multiple neurinomas and/or meningiomas in sporadic cases. (*1*) In contrast to the NF2 gene, deletion or loss-of-function mutation of the MN gene may be incompatible with differentiation or survival of germ cells or with embryonic development. (*2*) Homozygous deletion or loss-of-function mutation of the MN gene may be lethal in somatic cells. (*3*) Homozygous deletion or loss of the NF2 gene plus heterozygous loss of the MN gene may modify the type and/or grade of the tumor

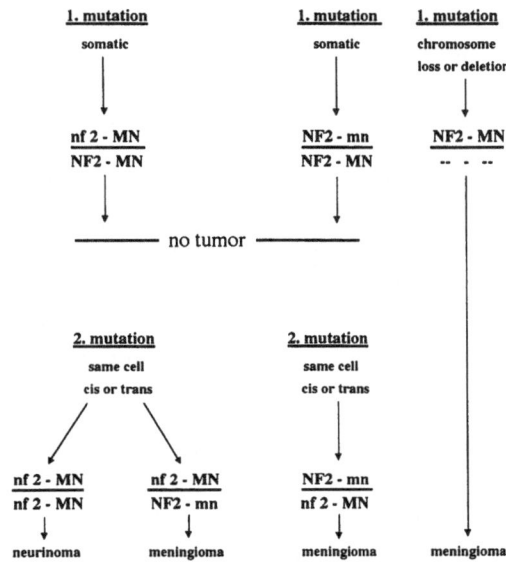

Fig. 3. Hypothetical molecular origin of sporadic neurinomas and meningiomas. (*1*) Neither heterozygous loss of the NF2 gene nor of the MN gene is sufficient to induce a tumor. (*2*) Homozygous loss of the NF2 gene induces a neurinoma. Heterozygous loss of both the NF2 gene and the MN gene induces meningioma

and as familial tumors with an autosomal dominant mode of inheritance (NF2).

– Patients with multiple neuromas and patients from neurofibromatosis families frequently also have meningiomas.

The three schematic drawings try to explain these uncommon epidemiologic and clinical findings. As a consequence, the following hypotheses have to be put forward:

1. A loss or a loss-of-function mutation of the NF2 gene is able to pass through the germline, unlike the deletion or loss-of-function mutation of the meningioma gene which seems to be a lethal factor during germ-cell differentiation or maturation or during early embryonic development. This would explain the fact that neuromas, but not meningiomas, can occur in a heritable familial form as NF2. If so, the origin of neuromas can be explained, in analogy to the retinoblastoma model, by two subsequent mutations of both alleles of the NF2 gene, with a homozygous loss of function of this gene.

   The occurrence of meningiomas, even multiple ones, could theoretically be explained analogously by first heterozygous, then homozygous loss of the meningioma gene in somatic meningothelial cells.

2. However, the occurrence of meningiomas in NF2 families, and especially the occurrence of meningiomas without accompanying neuromas in

family members, require additional modifying considerations which, moreover, would explain the significant frequency of the loss of a complete chromosome 22 in meningioma. Our hypothesis assumes that, in contrast to neuroma with homozygous loss of the NF2 gene, meningioma arises through combined heterozygous loss of one NF2 allele and one meningioma allele. This would allow a "one-hit" mechanism by loss of a complete chromosome 22 with a resulting double heterozygous gene loss. Furthermore, as shown in the three schemes, it would explain the considerably higher frequency of sporadic meningiomas than of neuromas.

Like many other hypothetical tumor models, this one too has its shortcomings. At the present state of molecular knowledge, it is unable to explain the fact that in about 30% of all meningiomas in vivo and in vitro, SV40 T antigen is detectable [8, 20]. One plausible explanation is that, like the RB and p53 proteins, the product of the regulatory gene or genes responsible for meningioma formation interacts with SV40 T antigen, with the result of a negative regulatory effect on the proliferation-suppressing function of the product of the meningioma gene.

Another open question is the effect of the loss of three alleles, which might occur under certain conditions. Does it lead to the loss of proliferation ability or death of the affected cell? Or, more likely, to tumor progression as is observed more frequently in meningiomas and neuromas of members of NF2 families?

These questions will not be answered until we have the sequences of both genes and their proteins in our hands and are able to investigate their functions.

References

1. Comings D (1973) A general theory of carcinogenesis. Proc Natl Acad Sci USA 70:3324–3328
2. Dumanski JP, Carlbom E, Collins VP, Nordenskjold M (1987) Deletion mapping of a locus on human chromosome 22 involved in the oncogenesis of meningioma. Proc Natl Acad Sci USA 84:9275–9279
3. Dumanski JP, Rouleau GA, Nordenskjöld M, Collins VP (1990) Molecular genetic analysis of chromosome 22 in 81 cases of meningioma. Cancer Res 50:5863–5867
4. Fearon ER, Vogelstein B (1990) A genetic model for colorectal carcinogenesis. Cell 61:759–767
5. Harris H (1988) The analysis of malignancy by cell fusion. Cancer Res 48: 3302–3306
6. Knudson AG Jr (1971) Mutation and cancer: statistical study of retinoblastoma. Proc Natl Acad Sci USA 68:820–823
7. Knudson AG Jr, Hethcote HW, Brown BW (1975) Mutation and childhood cancer. Proc Natl Acad Sci USA 72:5116–5120

8. Krieg P, Amtmann E, Jonas D, Fischer H, Zang KD, Sauer H (1981) Episomal simian virus 40 genomes in human brain tumors. Proc Natl Acad Sci USA 78:6446–6450
9. Krone W (1990) Modellbetrachtungen über Neurofibromatose 1 (von Recklinghausen). In: Imagines humanae. Ulmensien, Vol. 4. Ulm university press, Ulm, pp 65–79
10. Lekanne-Deprez RH, Groen NA, van Biezen NA, Hagemeijer A, van Drunen E, Koper JW, Avezaat CJ, Bootsma D, Zwarthoff EC (1991) A t(4; 22) in a meningioma points to the localization of a putative tumor-suppressor gene. Am J Hum Genet 48:783–790
11. Marshall CJ (1991) Tumor suppressor genes. Cell 64:313–326
12. Meese E, Blin N, Zang KD (1987) Loss of heterozygosity and the origin of meningioma. Hum Genet 77:349–351
13. Ohno S (1971) Genetic implications of karyological instability of malignant somatic cells. Physiol Rev 51:496–526
14. Peto R, Roe FJC, Lee PN, Levy L, Clack J (1975) Cancer and ageing in mice and man. Br J Cancer 32:411–426
15. Ponder BA (1988) Gene loss in human tumours. Nature 335:400–402
16. Rouleau GA, Seizinger BA, Wertelecki W, Haines JL, Superneau DW, Martuza RL, Gusella JF (1990) Flanking markers bracket the Neurofibromatosis 2 (NF2) gene on chromosome 22. Am J Hum Genet 46:323–328
17. Seizinger BR, De la Monte S, Atkin L, Gusella JF, Martuza RL (1987) Molecular genetic approach to human meningioma: Loss of genes on chromosome 22. Proc Natl Acad Sci USA 84:5419–5423
18. Seizinger BR, Martuza RL, Gusella JF (1986) Loss of genes on chromosome 22 in tumorigenesis of human acoustic neuroma. Nature 322:644–647
19. Stanbridge EJ, Nowell PC (1990) Origins of human cancer revisited – meeting review. Cell 63:867–874
20. Weiß AF, Portmann R, Fischer H, Simon J, Zang KD (1975) Simian virus 40 related antigens in three human meningiomas with defined chromosome loss. Proc Natl Acad Sci USA 72:609–613
21. Zang KD, Singer H (1967) Chromosomal constitution of meningiomas. Nature 216:84–85
22. Zankl H, Zang KD (1972) Cytological and cytogenetical studies on brain tumors IV. Hum Genet 14:167–169

p53: Oncogene, Tumor Suppressor, or Both?

W. Deppert

Heinrich-Pette-Institut für Experimentelle Virologie und Immunologie, Universität Hamburg, Martinistraße 52, W-2000 Hamburg 20, FRG

Oncogenes and Tumor Suppressor Genes in Human Cancer

Cytogenetic [34] and molecular genetic [25] analyses have identified tumor-specific genomic changes, thus providing accumulating evidence for a genetic component in the etiology of human cancer. Cancer results from multiple genetic lesions, induced by a variety of genotoxic agents. The ensuing mutations range from point mutations via deletions in a single gene up to chromosomal rearrangements. It can be expected that some of those mutations might activate genes that stimulate cell growth, while others might inactivate genes involved in negative regulation of cell growth. Until very recently cancer biologists in their overwhelming majority have focused on activated genes, i.e. the oncogenes. This, primarily, has experimental reasons. Oncogenes are mutant alleles of proto-oncogenes. Through their mutation they have acquired novel or aberrant properties that promote cell transformation. Oncogenes function in a genetically dominant manner and provide a selective growth advantage for cells expressing them, thus facilitating their identification [3, 4]. However, one has to be aware that genomic rearrangements are frequent events for mutations causally involved in the development of cancer, and are observed in virtually all tumor cells. One would expect that such large-scale genetic alterations will, with a higher frequency, result in elimination or inactivation rather than in activation of genes. Consequently, the inactivation of negatively regulating genes should comprise an important step in the development of cancer.

"Tumor suppressor genes", "anti-oncogenes", or "recessive oncogenes" are terms used to describe genes which can restrain tumor growth. Each of these terms describes a different aspect of the negative control of tumor growth, hence they are not freely interchangeable. A common denominator, however, is that elimination of their function is required for tumor growth. One can assume that multistep carcinogenesis requires both the action of activated oncogenes and the inactivation of tumor suppressor genes [4, 25].

Wagener/Neumann (Eds.)
Molecular Diagnostics of Cancer
© Springer-Verlag Berlin Heidelberg 1993

The prototype of a tumor suppressor gene is the retinoblastoma gene, which seems to be involved in negative regulation of cell growth by modulating the activity of the transcription factor E2F [35]. The list of tumor suppressor genes is rapidly growing, and comprises genes involved in diverse functions, ranging from transcription to cellular adhesion [2, 5].

p53 has been added to this list not too long ago. Meanwhile, it has been recognized as probably being one of the most important genes in the development of human cancer, as mutations in it reflect the most common genetic alteration in human cancer cells [6, 20]. Considering the history of p53, this rise was not at all to be expected.

From Tumor Antigen to Oncogene

p53 was discovered in 1979 as a cellular protein coprecipitating in immune complexes with the transforming protein of the small DNA tumor virus simian virus 40 (SV40), the SV40 tumor antigen (T-Ag) [26]. Later on, p53 was also found in many other nonvirally transformed tumor cells, but not in normal cells, and hence was termed a cellular tumor antigen. Improved methods of detection, however, demonstrated that p53 was also present in normal cells, albeit in grossly reduced levels. The functional role for p53, both in normal and in tumor cells, remained an enigma for many years (review in [8]). Around 1984, the p53 field started to gain momentum with two sets of observations:

1. Several laboratories demonstrated that the expression of p53 in normal cells is regulated in a cell cycle-dependent manner, suggesting a role for p53 in cellular proliferation. The view of p53 as a regulatory protein in the cell cycle was strongly supported by experiments performed by Mercer and colleagues [29], who showed that expression of p53 is required for the transit of resting cells from G_0 to G_1.
2. In a set of closely related experiments, a number of groups provided evidence for p53 being an oncogene: transfection of p53 into primary cells led to immortalization of these cells, while cotransfection of p53 with an activated *ras* oncogene resulted in conversion of the cells to a fully transformed phenotype. Furthermore, transfection of p53 expression vectors into a variety of cells indicated that overexpressed p53 induced effects related to cell transformation or tumorigenesis. In normal Rat-1 fibroblasts, p53 conferred a tumorigenic phenotype onto these cells; p53 enhanced the tumorigenic phenotype of a weakly tumorigenic Abelson murine leukemia virus-transformed cell, and it increased the metastatic potential of murine bladder carcinoma cells (review in [9]).

The effects of p53 in cell transformation and tumorigenesis, summarized in Table 1, were largely similar to the effects of *myc* overexpression, suggesting that p53 was a *myc*-like oncogene. Unlike with *myc*, however, overexpression

Table 1. Biological effects of transfected p53

Recipient cell	Effect
Primary rat embryo fibroblasts	Immortalization
Primary rat embryo fibroblasts	Phenotypic transformation in cooperation with activated *ras* oncogene
Adult rat chondrocytes	Immortalization
Mouse Swiss 3T3 cells	Abrogation of PDGF requirement
Abelson murine leukemia virus-transformed mouse L12 cell line	Conversion from a weakly to a highly tumorigenic phenotype
Murine bladder carcinoma cell line	Increase in metastatic potential
Rat-1 cell line	Enhancement of tumorigenicity without phenotypic transformation

The p53 proteins encoded were later found to be mutant proteins.

of p53 in tumor cells most often is not due to enhanced transcription, but to metabolic stabilization of the p53 protein at the post-translational level (review in [21]). This leads to transformed cells containing approximately 100-fold elevated levels of p53. As p53 in normal cells is required for cell cycle progression, it was assumed that the enhanced amounts of p53 in tumor cells would drive these cells from one round of the cell cycle into the next, i.e., would lead to enhanced proliferation (review in [8]). This view was further supported by the demonstration that overexpression of p53 could abrogate the PDGF requirement of nontransformed Swiss 3T3 cells [22] and, conversely, that elimination of p53 expression by vectors expressing p53 antisense RNA led to cessation of cell growth both in normal and in transformed cells [43].

From Oncogene to Tumor Suppressor

Later it was discovered that the p53 expression plasmids used in the studies described above had all encoded a mutant p53 (review in [32]). Repetition of these experiments with vectors encoding wild-type p53 then led to completely different results and to a completely different interpretation of the roles of p53 in cellular transformation and tumorigenesis. In cotransfection experiments with activated *ras*, wild-type p53 was completely devoid of any transforming activity [11, 15, 19]. This finding still could have been explained by assuming that wild-type p53, like many other proto-oncogenes, needed activating mutations to act as a dominant oncogene. However, further experiments cast doubts on the dominant oncogenic functions of even mutated p53. Wild-type p53 not only failed to immortalize primary cells, or to transform these cells in cooperation with *ras*, it even was able to suppress these events when cotransfected together with mutant p53 or the adenovirus E1A gene in such immortalization or transformation assays [12, 14]. Furthermore, transfection of wild-type p53 into p53-negative tumor cells led to a reversion of the tumorigenic phenotype and to a growth arrest of these cells at G_1 in the cell cycle [30]. Using temperature-sensitive mutants of p53, or wild-type p53 expressed under an inducible promoter, it was confirmed that, indeed, overexpression of wild-type p53 in tumor cells caused a reversible growth arrest, suggesting that wild-type p53 in such cells may have a negative effect on cell proliferation [31]. These experiments, together with observations described below, established wild-type p53 as a repressor of cell transformation and tumorigenesis. The properties of mutant and wild-type p53 with regard to cell transformation and tumorigenesis are compared in Table 2.

The classification of p53 as a tumor suppressor alleviated a long-standing problem in interpreting the discrepancy between the postulated oncogenic functions of p53 and the observation that in Friend virus-induced erythroleukemia in mice an appreciable number of cell clones had totally lost the ability to express p53, very often through severe rearrangements in the p53 gene. Those Friend cells still expressing p53 at all expressed a mutant p53 in 95% of cases [7, 36, 38]. The simplest interpretation of these data now was that loss of p53 function was an important step in the generation of Friend erythroleukemias and that introduction of a mutation into p53 rendered this protein nonfunctional, i.e., was equivalent to the loss of p53 expression.

The studies revealing the tumor suppressor functions of p53 in the rodent system greatly aided the studies of the role of p53 in human tumors, as it had been found that many human tumors carried mutations in the p53 gene. The p53 alterations observed ranged from total lack of p53 expression in some tumors to overexpression of a mutated p53 in others. Expression of a mutated p53 very often was accompanied by the loss of heterozygosity (LOH) (reviews in [28, 37]). LOH can be considered a property typical for tumor suppressor genes, indicating the total elimination of wild-type p53 function by mutational inactivation of one p53 allele coupled with the loss of the other allele. However, with increasing numbers of tumor cases analyzed,

Table 2. Functional comparison of properties of transfected wild-type and mutant p53 in cell transformation and tumorigenesis

Property analyzed	Wild-type p53	Mutant p53
Immortalization of primary cells	−	+
Cooperation with *ras* in phenotypic transformation	−	+
Repression of cell transformation (e.g., by E1A plus *ras*)	+	−
Reversion of tumorigenic phenotype in p53-negative tumor cells	+	−
Growth arrest in transformed cells	+	−

a growing number of tumors was found in which a mutant p53 allele was coexpressed together with the wild-type allele [39]. This set-up actually resembles the experimental situation encountered in cells transfected with a p53 mutant allele, e.g., in cells transformed with a mutant p53 gene and an activated *ras* oncogene. To explain the "dominance" of the mutant p53 allele, and to reconcile it with the recessive nature of tumor suppressor genes, it has been proposed that mutant p53 acts in a dominant-negative fashion by effectively blocking the function of the coexpressed wild-type p53 [17, 18]. Since mutant p53, due to its metabolic stability, is overexpressed compared to wild-type p53, a dominant-negative action of mutated p53 could be achieved by competing out cellular targets for wild-type p53. Alternatively, as p53 is an oligomeric protein, mutant p53 could form oligomers with wild-type p53, thus functionally eliminating wild-type p53. Indeed, hetero-oligomeric complexes between mutant and wild-type p53 have been described ([33], and literature cited therein).

However, one should also consider the possibility that mutant p53 can have an oncogenic function of its own, i.e., that a mutation in p53 might create a truly oncogenic p53, which has not merely lost a tumor suppressor function but actively contributes to the process of malignant transformation. In fact, the concept of "loss of function" as the single consequence for p53 mutations is already thrown into question by two long-known properties of mutant p53 listed in Table 1: Induction of a tumorigenic phenotype in a weakly tumorigenic Abelson murine leukemia-transformed cell line by transfected mutant p53 clearly demonstrated an oncogenic property of this mutant p53, since these cells did not express any endogenous p53. In addition, the increase of the metastatic potential of murine bladder carcinoma cells after transfection of mutant p53 strongly argued for a dominant-positive role of mutant p53 in tumorigenesis.

Characterization of p53 Mutations in Human Cancers

The establishment of a possibly dual effect of p53 mutations on p53 function was initiated by analyzing the wide spectrum of p53 mutations observed in human cancers and by comparing the properties of the various p53 mutants [6, 20]. The human p53 gene, located on chromosome 17p, has been analyzed in a wide variety of primary tumors, xenografts and cell lines derived from tumors. So far, p53 mutations have been found in virtually all cancer types looked at, with the interesting exception of a few neural tumors [6, 20]. p53 mutations in human tumors have a number of characteristic features: Firstly, most of them are missense point mutations giving rise to an altered protein. This implies that such an altered p53 had been selected during development of the tumor. Secondly, mutations in p53 are not randomly scattered over the protein, but cluster in certain areas of the protein. p53 contains five domains (I–V in Fig. 1), which are evolutionarily highly conserved among

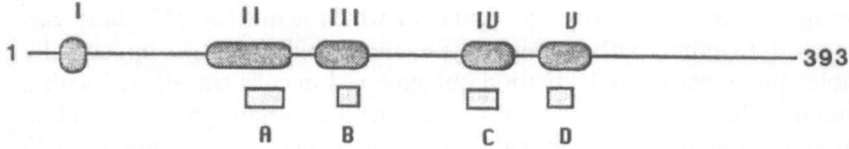

Fig. 1. Evolutionary conserved domains and mutational clusters in human p53

all species analyzed so far [44]. The majority of p53 mutations are clustered between amino acids 130 and 290 (out of 393). Within this area, most mutations are confined to regions II–V of the five conserved regions of the p53 protein (Fig. 1) [6]. The fact that mutations cluster within cross-species-conserved domains firstly suggests that they eliminate or change functional important domains of the p53 protein. However, the finding that different conserved domains are hit already suggests that mutations at different sites of the p53 gene may alter different functions of p53. The third important feature of p53 mutations is that there are at least three mutational "hot-spots", affecting codons 175, 248, and 273 (reviews in [6, 28]). The distribution of these "hot-spots" is quite different among different types of cancer. For example, mutation of codon 175 is not seen in lung tumors, although it is quite common in many other tumors, especially in colon carcinoma [6]. This is remarkable and could reflect the involvement of different mutagens or differences of other environmental aspects in the development of these tumors. Alternatively, different selective pressures for promoting cell growth or outgrowth of a tumorigenic cell might favor different mutational events in different tissues. The most striking example of specificity of p53 mutations was provided by the analysis of hepatocellular carcinoma (HCC) in persons from different geographical areas. A surprisingly high proportion (ca. 50%) of cases of HCC in areas of high incidence of infections with human hepatitis B virus (HBV), in China, or of nutritional exposure to aflotoxin, in Southern Africa, exhibited mutations at the third base pair position of codon 249 (most of them G to T transversions), leading to substitution of Arg by Ser in the mutant p53. In contrast, no codon 249 mutations were found in HCC from patients having a normal risk for this disease (reviews in [6, 20, 28]). Meanwhile, evidence for specificity in p53 mutations has also been provided by analysis of p53 in inherited forms of cancer, such as the Li-Fraumeni syndrome. This is a rare autosomal dominant syndrome, characterized by diverse neoplasms at many different sites of the body. Affected families have a high incidence of cancer. All of the families analyzed so far had p53 mutations clustering between codons 245 and 258, with approximately 50% of the mutations being at codon 248 (review in [6]). Although the molecular basis for this "heterospecificity" of p53 mutations is not yet understood, it strongly argues for the notion that different mutant p53 alleles might exhibit different molecular properties.

Properties of Wild-Type and Mutant p53

Unfortunately, characterization of p53 mutations is much more advanced at the level of the gene than at the protein level. Nevertheless, comparative analyses of wild-type p53 and several mutant p53 proteins both in the rodent and in the human system have permitted a list of properties distinguishing wild-type and mutant p53 to be compiled (Table 3). The most important one, and the one used in many diagnostic analyses, results from the fact that many – but by no means all! – mutant p53 proteins display an altered conformation which can be recognized by monoclonal antibodies specific for p53 in a mutant conformation [16]. Perhaps due to such a conformational alteration, wild-type and mutant p53 often differ in their interaction with cellular or viral target proteins. For example, while wild-type p53 forms a strong complex with SV40 large T, mutant p53 does not. Conversely, mutant p53 often strongly binds to the cellular 70-kDa heat shock proteins, a property not observed with wild-type p53 ([27] and literature cited therein). It is clear that these parameters can only provide a gross distinction between the wild-type and the mutant phenotype of p53 and that functional parameters are needed for further classification. This area of p53 research is just beginning, the main obstacle still being the absence of testable functions specific for wild-type p53. However, in looking for such functions, one has to be aware that "tumor suppression" is not a physiological function of p53 as such, but most likely reflects a control function of p53 in cellular proliferation or differentiation. Given the difficulties in defining such functions at the molecular level, the ensuing list of biological and biochemical properties compiled so far is impressive (Table 4). Not all of these activities have been tested both for wild-type p53 and for a major portion of the mutant p53. Certain prominent p53 mutations, however, have been analyzed in functional terms, and the results obtained so far clearly demonstrate that mutant p53 proteins still retain some of the activities of wild-type p53. Table 5 compares some selected properties of wild-type p53 with those of the three

Table 3. General properties of wild-type and mutant p53

Property analyzed	Wild-type p53	Mutant p53
Reactivity with monoclonal antibody		
PAb 1801	+	+
PAb 1620	+	−
PAb 240	−	+
Binding to SV40		
Large T antigen	+	−
Binding to cellular 70-kDa		
70 Kd heat shock protein	−	+
Half-life	~20 min	>3 hrs

Human p53 was used.

Table 4. Biological and biochemical properties of wild-type and mutant p53

Biological or biochemical activity analyzed	Wild-type p53	Mutant p53
Control of $G_0 \rightarrow G_1$ transit in cell cycle [29]	+	ND
Control of $G_1 \rightarrow S$ transit in cell cycle [10]	+	+
Transcriptional activation of test gene as GAL4 fusion protein [13, 41]	+	$\pm$[a]
Inhibition of SV40 T antigen-dependent SV40 DNA replication (reviews in [21, 28])		
In vivo (after transfection into CoS cells)	+	−
In vitro	+	−
DNA binding		
Nonspecific to ds-DNA [23, 45]	+	Reduced
Nonspecific to ss-DNA [23]	+	+
Sequence-specific to TGCCT, GCGGGG sequences [1, 24]	+	−
Sequence-/structure-specific to nuclear matrix attachment region (MAR) DNA [46]	+	+

ND, not determined.
[a] Strictly dependent on mutation.

p53 "hot-spot" mutants. It is evident that these mutants differ strikingly in their biological properties. The main feature is that a spectrum of mutations can be observed, from a mutant closely resembling wild type to mutants differing from wild type in all properties analyzed. However, one property seems to be common to all mutants analyzed so far: all of them have lost the "suppressor function" of p53, as defined by the ability to block cell growth after reintroduction into p53-negative tumor cells or to inhibit cell transformation in transfection assays with E1A and activated *ras* oncogenes.

Analysis of this phenomenon indicates that p53 mutations will fall into at least three different classes:

1. *True null mutations*. Such a p53 mutant will have lost its "suppressor function", but will score negative in all further assays for transforming activities, i.e. its presence is equivalent to the actual loss of the gene.
2. *Dominant-negative mutations*. In this class of mutants, the loss of "suppressor function" is coupled with the ability to interfere with the function of coexpressed wild-type p53. Mutants of this group will be recognized as oncogenic only in cells that coexpress a wild-type p53. It is important to realize that such mutants will not provide a direct oncogenic stimulus, but rather eliminate a (rate-limiting) inhibitory factor (i.e., the wild-type p53). With these mutants, the active transforming event must be provided by another positively dominant oncogene, and its action will be limited to provide loss of suppression.

Table 5. Properties of wild-type and "hot-spot" mutant p53

Property analyzed	Wild-type p53	Mutation of p53 at codon		
		175	248	273
Relative transformation frequency (in cooperation with *ras*)	0	3–10	ND	1
Suppression of transformation and growth of tumor cells	+	–	–	–
Conformational alterations				
Reactivity with mutant-specific monoclonal PAb 240	–	+	ND	–
Binding of 70-kDa heat shock protein	–	+	–	–
Binding of SV40 T antigen	+	–	ND	±
Transactivation of a test gene as GAL4 fusion protein	+	–	+	+
Half-life of protein	20 min	3–6 h	20 min	7 h

ND, not determined.

3. *Dominant-posiive mutations.* In addition to losing the "suppressor function", these mutations lead to the gain of an overt transforming function. Such p53 mutants should play a dominant-positive role in initiation or progression events during tumor formation. Such mutants also will actively contribute to growth deregulation and malignancy in cells that do not express endogenous p53.

Naturally occuring examples of all classes of mutants are found both in the human and in the rodent system, with clear classifications of null mutants, like the Li-Fraumeni codon 248 mutants, where the recessive nature of the mutation already is apparent from the history of the disease. Classifications of p53 mutations into the other two categories still depends on the experimental systems used, and it is conceivable that a mutant scoring "dominant-negative" in one assay system might score "dominant-positive" in another.

The high percentage of apparently "dominant-positive" p53 mutations, i.e., "gain-of-function" mutants, might also provide an answer to the intriguing question of why p53 mutations are so common in the development of human cancer: Most p53 mutations are point mutations, i.e. single hits. As all mutants analyzed so far have lost the p53 "suppressor function", an additional "gain of function" implies that a single point mutation would score as "two hits for one". Although this is an attractive hypothesis, the high frequency of p53 mutations might also be explained by assuming that alterations in the p53 gene represent an absolute restriction point ("bottleneck") in the development of a tumor.

Prospects for p53 in Tumor Diagnosis and Therapy

The high frequency of p53 mutations in human cancer render this protein an ideal tumor marker, and as such it is already used in many clinical studies. Mutant p53 in tumor cells is present in enhanced levels, facilitating its detection in tumor tissue by immunocytochemical methods. Furthermore, monoclonal antibodies with a prevalence for p53 in a mutant conformation exist (e.g., PAb 240) and are already providing some additional information on the conformational status of the p53 expressed. This is becoming increasingly important as cases accumulate where wild-type (or wild-type-like) p53 is found to be overexpressed in certain tumors [42]. However, p53 might provide more opportunities. Experiments have been initiated to probe the prognostic value of p53 analyses of various human tumors. Immunohistochemical studies seem to indicate that elevated p53 expression might be a marker for a more aggressive nature of the tumor [40]. So far, however, the prognostic power of p53 expression seems to be weak. The central issue will be whether enhanced p53 expression simply represents a random product of dedifferentiation or indeed is an important feature of a specific malignant phenotype, playing a key role in tumor behavior. This issue is not yet resolved, but considering the different nature of the p53 mutations and their functional consequences the current uncertainty is not surprising. Overexpression of p53 as such might be too unspecific a criterion for fine diagnosis of a tumor and thus for prognosis. Therefore, it will be necessary to characterize individual p53 mutations at the level of the gene and to correlate this information with diagnostic and prognostic parameters. Clearly, this is a major task, and even with the advancement of new PCR-based sequencing strategies this approach can hardly be used for routine diagnosis. However, such information might lead to the development of more refined, e.g., peptide-based, monoclonal antibodies detecting specific subclasses of p53 mutants. Such subclass-specific antibodies then would be applicable for routine diagnosis.

What about the therapeutic potential of p53? The finding that reintroduction of p53 into tumor cells can reverse the tumorigenic phenotype of p53-negative tumor cells prompted much optimism regarding the use of wild-type p53 expression vectors in successful somatic gene therapy. Indeed, the fact that wild-type p53 in such cells is able to overcome all adverse effects of other activated oncogenes, such as *ras* or *myc*, that are also expressed in these cells, is remarkable. Nevertheless, reasonable strategies for reintroduction and appropriate control of expression of wild-type p53 into tumors, especially solid tumors, are far away, and much more knowledge on the biology of p53 will be needed before such a strategy could be actively pursued. Even then, the most suitable and likely targets would be the rare hematopoetic tumors in which p53 mutations are found in stem cells. Nevertheless, these reintroduction studies clearly demonstrated that the tumorigenic phenotype of a cell can be permanently reversed. So the major therapeutic potential of these studies might be in the understanding of this reversal at the molecular level, perhaps allowing the design of drugs specifically mimicking the effect(s) of reintroduced wild-type p53.

Another strategic avenue for a p53-based therapeutic approach might result from the rediscovery that overexpressed, mutated p53 is a cellular tumor antigen. The fact that a certain proportion of sera from tumor patients display p53 antibodies shows that mutated (or even overexpressed) p53 is immunogenic. If it were possible to induce a cellular immune response against the mutated p53, strategies for immunological therapy of tumors displaying mutated p53 could be envisioned.

The ubiquitous nature and the high frequency of p53 mutations in human cancer strongly suggest that p53 is one of the key molecules in the development of human cancer. Therefore, a better understanding of the functions of p53, wild type and mutant, at the molecular level not only will further our knowledge about the biological role of this enigmatic protein but also holds great promise for improvement of both cancer diagnosis and cancer therapy.

References

1. Bargonetti J, Friedman PN, Kern SE, Vogelstein B, Prives C (1991) Wild-type but not mutant p53 immunopurified proteins bind to sequences adjacent to the SV40 origin of replication. Cell 65:1083–1091
2. Birchmeier W, Behrens J, Weidner KM, Frixen UH, Schipper J (1991) Dominant and recessive genes involved in tumor cell invasion. Curr Opin Cell Biol 3:832–840
3. Bishop JM (1987) The molecular genetics of cancer. Science 235:305–311
4. Bishop JM (1991) Molecular themes in oncogenesis. Cell 64:235–248
5. Boyd JA, Barrett JC (1990) Tumor suppressor genes: possible functions in the negative regulation of cell proliferation. Mol Carcinog 3:325–329
6. Caron de Fromentel C, Soussi T (1992) The TP53 tumor suppressor gene: a model for investigating human mutagenesis. Genes, Chromosome Cancer 4:1–15
7. Chow V, Ben-David Y, Bernstein A, Benchimol S, Mowat M (1987) Multistage Friend erythroleukemia: independent origin of tumor clones with normal or rearranged p53 cellular oncogenes. J Virol 61:2777–2781
8. Crawford L (1983) The 53,000-dalton cellular protein and its role in transformation. Int Rev Exp Pathol 25:1–50
9. Deppert W (1989) P53: oncogene or anti-oncogene? A critical review. In: Lother H, Dernick R, Ostertag W (eds) Vectors as tools for the study of normal and abnormal growth and differentiation. Springer, Berlin Heidelberg New York, pp 399–406 (NATO ASI series, vol H34)
10. Deppert W, Buschhausen-Denker G, Patschinsky T, Steinmeyer K (1990) Cell cycle control of p53 in normal (3T3) or chemically transformed (Meth A) mouse cells: II. Requirement for cell cycle progression. Oncogene 5:1701–1706
11. Eliyahu D, Goldfinger N, Pinhasi-Kimhi O, Shaulsky G, Skurnik Y, Arai N, Rotter V, Oren M (1988) Meth A fibrosarcoma cells express two transforming mutant p53 species. Oncogene 3:313–321
12. Eliyahu D, Michalovitz D, Eliyahu S, Pinhasi-Kimhi O, Oren M (1989) Wild-type p53 can inhibit oncogene-mediated focus formation. Proc Natl Acad Sci USA 86:8763–8767

13. Fields S, Jang SK (1990) Presence of a potent transcription activating sequence in the p53 protein. Science 249:1046–1049
14. Finlay CA, Hinds PW, Levine AJ (1989) The p53 proto-oncogene can act as a suppressor of transformation. Cell 57:1083–1093
15. Finlay CA, Hinds PW, Tan T-H, Eliyahu D, Oren M, Levine AJ (1988) Activating mutations for transformation by p53 produce a gene product that forms an hsc70-p53 complex with an altered half-life. Mol Cell Biol 8:531–539
16. Gannon JV, Greaves R, Iggo R, Lane DP (1990) Activating mutations in p53 produce a common conformational effect. A monoclonal antibody specific for the mutant form. EMBO J 9:1595–1602
17. Green MR (1989) When the products of oncogenes and anti-oncogenes meet. Cell 56:1–3
18. Herskowitz I (1987) Functional inactivation of genes by dominant negative mutations. Nature 329:219–222
19. Hinds P, Finlay C, Levine AJ (1989) Mutation is required to activate the p53 gene for cooperation with the ras oncogene and transformation. J Virol 63: 739–746
20. Hollstein M, Sidransky D, Vogelstein B, Harris CC (1991) p53 Mutations in human cancers. Science 253:49–53
21. Jenkins JR, Stürzbecher H-W (1988) The p53 oncogene. In: Reddy EP, Skalka AM, Curran T (eds) The oncogene handbook. Elsevier, Amsterdam
22. Kaczmarek L, Oren M, Baserga R (1986) Co-operation between the p53 protein tumor antigen and platelet-poor plasma in the induction of cellular DNA synthesis. Exp Cell Res 162:268–272
23. Kern SE, Kinzler KW, Baker SJ, Nigro JM, Rotter V, Levine AJ, Friedman P, Prives C, Vogelstein B (1991) Mutant p53 proteins bind DNA abnormally in vitro. Oncogene 6:131–136
24. Kern SE, Kinzler KW, Bruskin A, Jarosz D, Friedman P, Prives C, Vogelstein B (1991) Identification of p53 as a sequence-specific DNA-binding protein. Science 252:1708–1711
25. Klein G (1987) The approaching era of the tumor suppressor genes. Science 238:1539–1545
26. Lane DP, Crawford LV (1979) T antigen is bound to a host protein in SV40-transformed cells. Nature 278:261–263
27. Lehman TA, Bennett WP, Metcalf RA, Welsh JA, Ecker J, Modali RV, Ullrich S, Romano JW, Appella E et al. (1991) p53 Mutations, ras mutations, and p53-heat shock 70 protein complexes in human lung carcinoma cell lines. Cancer Res 51:4090–4096
28. Levine AJ, Momand J, Finlay CA (1991) The p53 tumour suppressor gene. Nature 351:453–456
29. Mercer WE, Avignolo C, Baserga R (1984) Role of p53 protein in cell proliferation as studied by microinjection of monoclonal antibodies. Mol Cell Biol 4:276–281
30. Mercer WE, Shields MT, Amin M, Sauve GJ, Appella E, Romano JW, Ullrich SJ (1990) Negative growth regulation in a glioblastoma tumor cell line that conditionally expresses human wild-type p53. Proc Natl Acad Sci USA 87: 6166–6170
31. Michalovitz D, Halevy O, Oren M (1990) Conditional inhibition of transformation and of cell proliferation by a temperature-sensitive mutant of p53. Cell 62:671–680
32. Michalovitz D, Halevy O, Oren M (1991) p53 Mutations: gains or losses? J Cell Biochem 45:22–29
33. Milner J, Medcalf EA (1991) Cotranslation of activated mutant p53 with wild type drives the wild-type p53 protein into the mutant conformation. Cell 65: 765–774

34. Mitelman F (1988) Catalog of chromosome aberrations in cancer. Liss, New York
35. Moran E (1991) Cycles within cycles. Curr Op Cell Biol 1:281–283
36. Mowat M, Cheng A, Kimura N, Bernstein A, Benchimol S (1985) Rearrangements of the cellular p53 gene in erythroleukaemic cells transformed by Friend virus. Nature 314:633–636
37. Mulligan LM, Matlashewski GJ, Scrable HJ, Cavenee WK (1990) Mechanisms of p53 loss in human sarcomas. Proc Natl Acad Sci USA 87:5863–5867
38. Munroe DG, Rovinski B, Bernstein A, Benchimol S (1988) Loss of a highly conserved domain on p53 as a result of gene deletion during Friend virus-induced erythroleukemia. Oncogene 2:621–624
39. Nigro JM, Baker SJ, Preisinger AC, Jessup JM, Hostetter R, Cleary K, Bigner SH, Davidson N, Baylin S, Devilee P, Glover T, Collins FS, Weston A, Modali R, Harris CC, Vogelstein B (1989) Mutations in the p53 gene occur in diverse human tumour types. Nature 342:705–708
40. Ostrowski JL, Sawan A, Henry L, Wright C, Henry JA, Hennessy C, Lennard TJW, Angus B, Horne CHW (1991) p53 Expression in human breast cancer related to survival and prognostic factors: an immunohistochemical study. J Pathol 164:75–81
41. Raycroft L, Wu H, Lozano G (1990) Transcriptional activation by wild-type but not transforming mutants of the p53 anti-oncogene. Science 249:1049–1051
42. Rodrigues NR, Rowan A, Smith MEF, Kerr IB, Bodmer WF, Gannon JV, Lane DP (1990) p53 Mutations in colorectal cancer. Proc Natl Acad Sci USA 87:7555–7559
43. Shohat O, Greenberg M, Reisman D, Oren M, Rotter V (1987) Inhibition of cell growth mediated by plasmids encoding p53 anti-sense. Oncogene 1:277–283
44. Soussi T, Caron de Fromentel C, May P (1990) Structural aspects of the p53 protein in relation to gene evolution. Oncogene 5:945–952
45. Steinmeyer K, Deppert W (1988) DNA binding properties of murine p53. Oncogene 3:501–507
46. Weißker S, Müller B, Homfeld A, Deppert W. (1992) Specific and complex interactions of murine p53 with DNA. Oncogene 7:1921–1932

Tumor Genome Screening by Multilocus DNA Fingerprints as Obtained by Simple Repetitive Oligonucleotide Probes

J.T. Epplen,[1] S. Bock,[2] and P. Nürnberg[3]

1 Molekulare Humangenetik, Medizinische Fakultät, Ruhr-Universität,
  Universitätsstraße 150, W-4630 Bochum, FRG
2 Institut für klinische Hämatologie der GSF, Forschungszentrum für Umwelt
  und Gesundheit GmbH, Hämatologikum, W-8000 München 70, FRG
3 Institut für Medizinische Genetik des Bereichs Medizin (Charité)
  Humboldt-Universität Berlin, 0-1040 Berlin, FRG

Introduction

The process of neoplastic transformation and its sequelae are generally accompanied by profound genomic alterations. Gain or loss of whole chromosomes and gross chromosomal translocations or other rearrangements are readily detectable by cytogenetic investigation [31]. In order to detect minor somatic changes such as microdeletions or point mutations, direct DNA analyses are necessary [39]. Specific oncogene amplification events can be ascertained via, for example, the Southern blot hybridization technique [4]. Cytogenetic analyses have revealed a multitude of different numerical and structural abnormalities in gliomas of high and low grades of malignancy [5, 15, 18]. The most common features are early loss of gonosomes and gain of additional copies of chromosome 7. The most frequent aberration in human gliomas is the presence of double minutes (dmin) in nearly 50% of the cases with higher malignancy grade [3, 4]. Double minutes are cytogenetic equivalents of gene amplification [11] and in human gliomas they usually coincide with amplification of the epidermal growth factor receptor (EGFR) gene [4]. Recently DNA fingerprinting has been applied to detect somatic changes in human cancer DNA [25, 28]. As in karyotyping, oligonucleotide fingerprinting allows a simultaneous survey of the alterations in the whole genome for numerous highly variable DNA loci which are scattered over the chromosomes [7]. Therefore fingerprinting may serve to supplement or sometimes even replace karyotype analysis, which is technically difficult in solid tumors and impossible when only frozen tissue is available.

In the context of cancer research we have used simple repetitive oligonucleotide probes for various aspects of DNA fingerprinting as first described by Ali et al. [1] for the following reasons: (a) In-gel hybridization with synthetic oligonucleotides is technically superior to other probes [26] and therefore better suited for routine scanning of human tumor DNA with

Wagener/Neumann (Eds.)
Molecular Diagnostics of Cancer
© Springer-Verlag Berlin Heidelberg 1993

general diagnostic intention, such as assessment of tumor homogeneity and progression [10, 19]. (b) The most informative (cac/gtg)$_n$ loci have been shown to coincide with R-band regions on human metaphase chromosomes [42], suggesting distribution over the whole genome, whereas minisatellite structures [16, 17] cluster predominantly in telomeric regions [30]. (c) Multiple rehybridizations with the whole panel of oligonucleotide probes are conveniently performed in one gel requiring only comparatively small amounts of DNA (and tumor material).

Materials and Methods

Isolation of DNA from cells and tissue was performed according to standard procedures [32]. A novel "quick preparation" procedure (<1 h duration before restriction enzyme digestion; Genomix, Kontron) proved very efficient for DNA extraction. This method has already been used successfully for DNA preparation from many tissue sources. Especially extractions from peripheral blood are performed on a routine basis by this method in our laboratory. About 6 µg from each DNA sample was digested with the restriction endonucleases *MboI*, *HinfI* and *HaeIII* according to the suppliers' recommendations. Whenever possible, digestion of DNA was performed in the absence and in the presence of 10 mM spermidine, in an additional aliquot, to exclude "hidden partial" digestions. The DNA fragments were separated on 0.7% horizontal agarose gels in TAE buffer for up to 48 h at 1 V/cm [8]. ^{32}P-labeling of the oligonucleotide probes as well as hybridization in the gel were performed as described previously by Schäfer et al. [33].

For Southern blot analyses 6 µg DNA was digested with the restriction endonucleases *PstI*, *EcoRI* and *HaeIII*. Electrophoresis was performed as described above, but 1 × TBE buffer was used instead. Separated DNA fragments were transferred onto nylon membranes according to standard procedures [32]. A full-length EGFR cDNA probe included, in addition to the coding sequence, both 5'- and extended 3'-untranslated sequences. It was labeled using the multipriming ([α^{32}P]dCTP + [α^{32}P]dATP). Hybridization and washing was performed under conditions of high stringency. After autoradiography the EGFR cDNA probe was stripped off and the blots rehybridized successively to (GT)$_8$ and (GTG)$_5$ as described above.

The detailed description of the teratocarcinoma cell lines, their history and their various culturing conditions has been the subject of a previous report [35].

Results

The Informativeness of DNA Fingerprints Compared with Tumor Cytogenetics in Gliomas

DNA fingerprints from the pathological tissue and/or primary cell culture of 80 intracranial tumors (mostly gliomas and a few medulloblastomas and metastases of carcinomas) were generated by the oligonucleotide probe $(GTG)_5$ and compared with the constitutional band pattern obtained from the peripheral blood leukocytes (PBL) of each patient. In parallel, karyotype analysis was performed from primary cultures of tumor cells and PBL of every patient but one. Numerous differences between tumor and constitutional DNA fingerprints were found [27]. According to our previous results on the complete somatic stability of $(cac/gtg)_n$ loci including different areas of the human brain [26], these changes are obviously a consequence of the malignant transformation, but not due to tissue-specific genomic rearrangements [28a]. Band intensity shifts clearly prevailed, but complete loss of bands as well as gain of new bands was also observed. Many changes present in the original tumor escaped detection when only primary cell cultures were investigated. Using a single hybridization probe a detection rate of nearly 80% for glioma-associated somatic changes in the DNA was reached. The parallel analysis employing three different restriction enzymes was useful but not decisive. Only three gliomas showed no deviation in their fingerprints after cleavage with *Mbo*I but did show deviation using *Hin*fI or *Hae*III. Generally each enzyme was as informative as the other two with respect to the number of somatic changes detected [27]. The direct comparison of cytogenetic and fingerprint data revealed that only one patient showed a detectable alteration in the tumor-derived DNA fingerprint without any chromosomal deviation. On the other hand, several chromosomal aberrations (loss of Y, X or 15 or gain of 7 or dmin) escaped from the detection by DNA fingerprinting. Two tumors with extensive chromosomal abnormalities also failed to show deviations in DNA fingerprint analysis. This seems to be due to the fact that only cultured cells and not the tumor itself could be analyzed. In a subset of 26 of the first 35 tumors investigated, the altered karyotypes correlated well with deviations in the DNA fingerprints, though sometimes loss of chromosomes was accompanied by gain of bands or vice versa. Interestingly, there was a coincidence of dmin with the prominent new fingerprint fragments described above. Only one tumor exhibiting these fragments had no demonstrable dmin in its karyotype.

Amplification Events and Telomere Lengths in Gliomas

Initially the most striking alterations in the tumor DNA fingerprints were found in the low-molecular-weight region of the gels (see Fig. 1, arrows

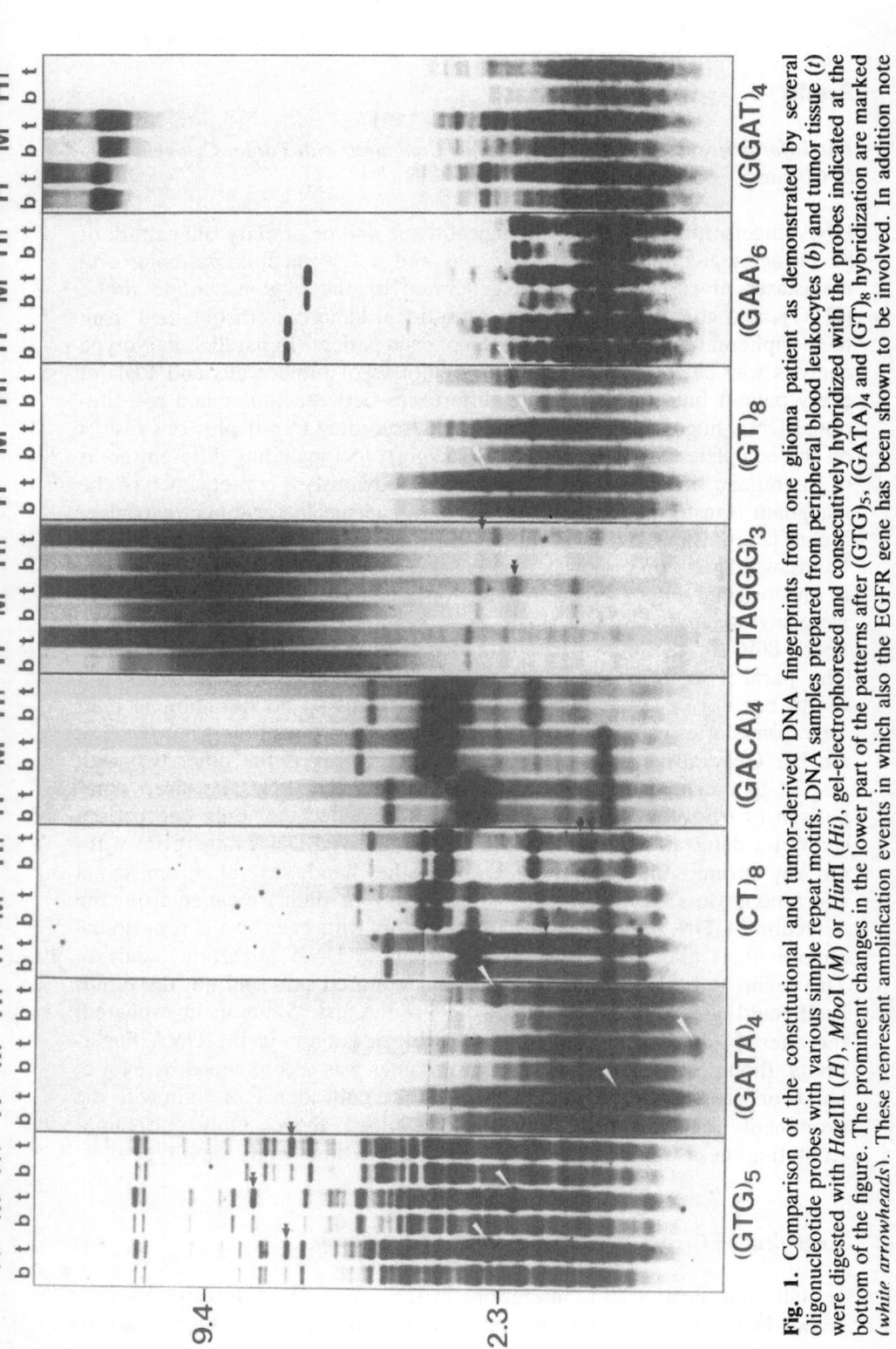

Fig. 1. Comparison of the constitutional and tumor-derived DNA fingerprints from one glioma patient as demonstrated by several oligonucleotide probes with various simple repeat motifs. DNA samples prepared from peripheral blood leukocytes (*b*) and tumor tissue (*t*) were digested with *Hae*III (*H*), *Mbo*I (*M*) or *Hinf*I (*Hi*), gel-electrophoresed and consecutively hybridized with the probes indicated at the bottom of the figure. The prominent changes in the lower part of the patterns after (GTG)₅, (GATA)₄ and (GT)₈ hybridization are marked (*white arrowheads*). These represent amplification events in which also the EGFR gene has been shown to be involved. In addition note

at the bottom). Eight of 80 unrelated glioma DNAs (10%) showed a prominent new or at least dramatically intensified band after restriction with *Hae*III (2.4 kb) and *Mbo*I (2.1 kb). Equivalent new bands could not be detected after *Hin*fI digestion. At least three patients exhibited additional prominent tumor-specific bands in all or some digests. Their size ranged from 1.2 to 1.5 kb depending on the restriction enzyme used. Rehybridization of the gels with $(GT)_8$ revealed similar striking bands in each of these eight patients (2.6 kb after *Hae*III, 2.0 kb after *Mbo*I and 2.2 kb after *Hin*fI; Fig. 1). Furthermore, intense signal bands were also present in the $(GATA)_4$ fingerprint of one glioma (*Hae*III 1.4 kb, *Mbo*I 1.2 kb, *Hin*fI 2.4 kb) but absent in the other tumors (Fig. 1). Irrespective of the probe employed, such prominent novel fragments could not be detected in a medulloblastoma and four metastasic carcinomas also included in this study. This supports the idea of a glioma-specific phenomenon. A correlation between the malignancy grade of the gliomas studied and the appearance of the prominent new fragments was not immediately obvious. But there does exist a remarkable correlation between malignancy grade and the probability to detect genomic deviations by DNA fingerprinting (with only two exceptions).

In addition to the probes mentioned above, we also hybridized the gels with the telomere-derived simple repetitive oligonucleotide $(TTAGGG)_3$. This probe provides cumulative information about the lengths and total number of telomeric sequences as well as about the organization of sequence-related, interspersed, simple repetitive structures. The simultaneous evaluation showed that 37% of the gliomas ($n = 33$) did not change the overall lengths of the telomeric sequences in comparison with the peripheral blood DNA, another 37% had increased telomeres, while in the remaining 26% of gliomas tumor telomeres had grown shorter [28a]. The necessary comparisons with normal glial DNA are currently being performed. Yet the tendency towards longer telomere structures stands in contrast to a report by Hastie et al. [14], who described the telomeres in colorectal tumors as significantly shorter than in the respective normal tissue or sperm.

In addition to the fingerprint analyses a collection of 15 gliomas was tested for amplification of the EGFR gene. Amplification of this sequence often accompanies the tumorogenesis in glial tissue [4, 22, 41]. In precisely the aforementioned five tumors showing the amplified fingerprint fragments, a dramatic elevation of the EGFR gene copy number was found. Moreover, rehybridization of the Southern blots with the relevant simple repeat oligonucleotides revealed identical positions for the largest *Hae*III fragment (2.9 kb), the fourth largest *Pst*I fragment (3.0 kb) and the third largest *Eco*RI fragment (5.9 kb) and the corresponding prominent $(GT)_8$ fingerprint bands [28]. No comigration was detected between any EGFR exon-bearing fragment and the $(GTG)_5$ fingerprint band signals that were drastically increased in intensity. This was already obvious from the fact that the $(GTG)_5$ fragments varied in size whereas the former did not.

DNA Fingerprinting for the Identification of Cell Lines in Tumor Research

Whenever cultured cells derived from different individuals have to be differentiated DNA fingerprints offer this possibility conveniently. For example, when in immunological studies T lymphocytes are stimulated specifically in culture, then nonproliferating antigen-presenting cells have to be present. Upon insufficient X-ray or antimetabolic treatment such feeder cells may inadvertently overgrow the original cells. Thus such cell culture procedures must be permanently controlled appropriately [13a, 22a]. In addition, the myeloma partners for the fusion with antibody-producing B lymphocytes can be differentiated by their fingerprint patterns as well as the derived fusion products secreting various immunoglobulins [13a]. Finally, therefore, oligonucleotide fingerprinting was also applied to investigate the relatedness of several cell lines that were established between 1973 and 1977 from a teratocarcinoma [35]. We were able to distinguish cell lines derived at different times. In addition, sublines from one cell line could be discriminated by using a combination of different probes. Therefore, multilocus fingerprinting with oligonucleotides is a useful method for monitoring changes in cell lines kept in culture for many generations.

Discussion

The analysis of somatic changes in the DNA of tumors is one more key to a comprehensive understanding of cancerogenesis and tumor progression [9] and more specifically of gliomas [3]. We would like to stress that a high number of quantitative as well as qualitative deviations can be detected in the DNA of human gliomas by using a single fingerprint probe, namely the simple repetitive oligonucleotide $(GTG)_5$. Taking into account that tumor cells are often overgrown by normal diploid cells in primary cell cultures, as shown by Teyssier [37], our data based solely on native tumor tissue indicate a detection rate of somatic changes in a very high proportion (78%) of gliomas. This is the more surprising as we carefully avoided scoring any artifacts in the banding patterns (e.g., "hidden partial" digestions [24]). Particular bands show shifts in their positions when compared to standards. Apparently certain restriction sites are less accessible or sensitive to cutting activity than the bulk of recognition sites. This serious problem has not been dealt with in other work reporting lower percentages for tumor-associated changes in the DNA fingerprints [10, 19, 21, 29, 34, 36]. Interestingly, previous investigators used combinations of up to three different fingerprint probes, while here only $(GTG)_5$ was employed. This underscores the discrepancies in detection rates. Indeed, successive hybridization of the gels with a whole panel of various simple repeat probes allows identification of additional differences between constitutional and tumor DNA, thereby increasing the overall detection rate to some 85% [27]. The type of tumor investigated is not the reason for the high percentage of deviations found, since the detection rate was not lower in the nonglioma tumors included in this study. Rather, the type of probe used and methodological refinements in high-resolution fingerprints account for the contrasting results. The finger-print changes in tumor relapses correlate quite well with the pathological and anatomical staging: If the novel tumor is classified in a more malignant category, then the fingerprint deviations tend to increase.

A number of different processes may be involved in the alteration of the DNA fingerprint in the tumor. Parallel karyotype analyses support the idea that chromosomal abnormalities are a major cause for deviations in the DNA fingerprints: Gain and loss of chromosomes or chromosomal regions obviously resulted (a) in intensity shifts of associated fingerprint fragments (due to monosomies or trisomies in parts of the tissue), (b) in novel tumor bands or (c) in loss of constitutional bands. But discrepancies between fingerprint and karyotype alterations, found in several cases, reveal that other processes are also operating. Hypomethylation of tumor DNA has been shown to occur in colon neoplasms [12]. This could also affect DNA fingerprints [10, 38]: using the methylation-sensitive restriction enzyme $HinfI$, the predominant occurrence of "hidden partials" in leukocyte DNA is explainable (GANT^{m}C is cleaved with detectable rate differences, decreasing in the order unmethylated, hemimethylated and bimethylated; G^{m6}ANTC and GANT^{hm5}C are not cleaved [23]). The novel bands detected in seven gliomas may be caused by generation of new length alleles at particular loci in addition to chromosomal rearrangements [2]. In principle they could have arisen via unequal sister-chromatid exchange during mitosis or, less likely, by replication slippage or insertion/deletion of repeat units as hypothesized by Wolff and coworkers [40]. The aforementioned mechanisms could also account for the loss of bands observed in six gliomas.

Strong novel (or at least drastically intensified) bands in tumors are perhaps due to localized amplification of DNA sequences as supposed by Stark and Wahl [36]. Such events may well be the reason for the appearance of the prominent fingerprint fragments found here in five gliomas. The concept of amplification is supported by the appearance of dmin in 80% of the cases and by the parallel amplification of the EGFR gene detected in all cases. Amplification units have been reported to cover up to several hundred kilobases [20]. Until recently it has not been known whether the 110-kb EGFR locus [13] harbors simple repeat loci such as $(ca/gt)_n$ or $(cac/gtg)_n$. On the basis of the uniform shift observed for the amplified fingerprint fragments using different restriction enzymes, the loci affected in the various patients are apparently identical. Accepting the concept of amplification, one has to explain why the $(cac/gtg)_n$ fragments in some tumors appeared as novel bands. The reason could be a short stretch of simple repeats on the respective fragments not detectable as single copy in the fingerprint prior to amplification. Alternatively, an altered position of the fragment in the fingerprint as a consequence of the amplification is plausible. The latter would be the case if the simple repetitive triplet structure were located on the edge of the amplification unit and one of the relevant restriction sites were not included. The length of the amplification unit, however, does not seem to be constant, since an additional $(cac/gtg)_n$-containing fragment was coamplified only in a subset of the tumors.

The comigration observed after HaeIII, PstI and EcoRI digestion of one amplified EGFR gene fragment and the prominent $(ca/gt)_n$ band suggests intragenic localization of the simple repeat. According to the published EcoRI restriction map of the EGFR locus [13], the $(ca/gt)_n$ repeat is probably situated on the fragment bearing the exons 5–7. So far, EGFR gene amplification has only been found to occur in gliomas (and some squamous cell carcinomas) [6].

In conclusion, the high detection rate of clonal markers in tumor DNA via DNA fingerprinting with the synthetic oligonucleotide $(GTG)_5$, taken together with its easy handling, favors this probe for routine applications. A correlation, albeit not absolute, between malignancy grade and appearance of deviations in the DNA fingerprint is evident. In cases where karyotype analysis is not feasible and/or fast analysis is necessary, DNA fingerprinting is a sound alternative or supplement to conventional methods. Moreover, this technique provides additional unique information about genomic alterations not obtainable with other methods. In the end it may even enable us to discover novel DNA sequences involved in the cellular transformation process.

Summary

Multilocus DNA fingerprints were generated by oligonucleotide probes specific for simple repeats from surgically removed tissue and/or cultured cells of intracranial tumors (gliomas, medulloblastoma, metastatic carcinomas) as well as other carcinomas. These were compared with the constitutional banding patterns obtained from peripheral blood leukocytes or normal tissue adjacent to the tumor of each patient. The probe $(CAC)_5$ or its complement $(GTG)_5$ proved especially efficient at revealing randomly distributed, unknown genomic alterations in human tumors. Prominent changes were observed also with other simple repeat oligonucleotides, e.g., in most renal tumors investigated [5a]. A multitude of somatic changes were detected and found to reflect the chromosome alterations identified by parallel karyotype analysis in the gliomas. Gain and/or loss of bands or significant band intensity shifts could be demonstrated in the fingerprints of more than 80% of the intracranial tumors investigated. This included highly amplified DNA fingerprint fragments in independent gliomas. Amplification of the epidermal growth factor receptor (EGFR) gene via Southern blot hybridization was revealed only in those tumors showing amplified DNA fingerprint fragments as well. In contrast to the situation reported for other tumors, the telomere lengths in gliomas tend to be increased rather than decreased in comparison with peripheral blood DNA. In a collateral approach, long-term cultured cell lines from a mouse teratocarcinoma were efficiently monitored by multilocus fingerprinting. In addition, particular sublines from one cell line of this tumor can be differentiated easily by a combination of simple repeat oligonucleotide probes.

Acknowledgments. This work was supported by the Volkswagen-Stiftung. The multilocus oligonucleotides and hypervariable single locus probes are subject to patent applications; commercial enquiries should be directed to Fresenius AG, Oberursel, FRG.

References

1. Ali S, Müller CR, Epplen JT (1986) DNA fingerprinting by oligonucleotides specific for simple repeats. Hum Genet 74:239–243
2. Armour JAL, Patel I, Thein SL, Fey MF, Jeffreys AJ (1989) Analysis of somatic mutations at human minisatellite loci in tumors and cell lines. Genomics 4: 328–334
3. Bigner SH, Vogelstein B (1990) Cytogenetics and molecular genetics of malignant gliomas and medulloblastomas. Brain Pathol 1:12–18
4. Bigner SH, Wong AJ, Mark J, Muhlbaier LH, Kinzler KW, Vogelstein B, Bigner DD (1987) Relationship between gene amplification and chromosomal deviations in malignant human gliomas. Cancer Genet Cytogenet 29:165–170
5. Bigner SH, Mark J, Burger PC, Mahaley MS Jr, Bullard DE, Muhlbaier LH, Bigner DD (1988) Specific chromosomal abnormalities in malignant human gliomas. Cancer Res 48:405–411
5a. Bock S, Epplen JT, Noll-Puchta H, Rotter M, Höfler H, Block T, Hartung R, Jakse G, Wilmanns W, Petrides PE (1992) Detection of somatic changes in human renal cell carcinomas with oligonucleotide probes specific for simple repeat motifs. Genes Chromosomes Cancer (in press)
6. Burck KB, Liu ET, Larrick JW (1988) Oncogenes: an introduction to the concept of cancer genes. Springer, Berlin Heidelberg New York
7. Epplen JT, Ammer H, Epplen C, Kammerbauer C, Roewer L, Schwaiger W, Steimle V, Zischler H, Albert E, Andreas A, Beyermann B, Meyer W, Buitkamp J, Nanda I, Schmid M, Nürnberg P, Pena SDJ, Pöche H, Sprecher W, Schartl M, Yassouridis A (1991) Oligonucleotide fingerprinting using simple repeat motifs: a convenient, ubiquitously applicable method to detect hyper-variability for multiple purposes. In: Burke T, Dolf G, Jeffreys AJ, Wolff R (eds) DNA fingerprinting: approaches and applications. Birkhäuser, Basel, pp 50–69
8. Epplen JT (1991) The methodology of multilocus DNA fingerprinting using radioactive or non-radioactive oligonucleotide probes specific for simple repetitive motifs. In: Chrambach A, Dunn MJ, Radola BJ (eds) Advances in electrophoresis, vol 5. VCH, Weinheim, pp 59–114
9. Fearon ER, Vogelstein B (1990) A genetic model for colorectal tumorigenesis. Cell 61:759–767
10. Fey MF, Wells RA, Wainscoat JS, Thein SL (1988) Assessment of clonality in gastrointestinal cancer by DNA fingerprinting. J Clin Invest 82:1532–1537

11. Gebhart E, Brüderlein S, Tulusan AH, von Maillot K, Birkmann J (1984) Incidence of double minutes, cytogenetic equivalents of gene amplification, in human carcinoma cells. Int J Cancer 34:369–373

12. Goelz SE, Vogelstein B, Hamilton SR, Feinberg AP (1985) Hypomethylation of DNA from benign and malignant colon neoplasms. Science 228:187–190

13. Haley J, Whittle N, Bennett P, Kinchington D, Ullrich A, Waterfield M (1987) The human EGF receptor gene: structure of the 110 kb locus and identification of sequences regulating its transcription. Oncogene Res 1:375–396

13a. Hampe J, Nürnberg P, Epplen C, Jahn S, Grunow R, Epplen JT (1992) Oligomucleotide fingerprinting as a means to identify and survey long-term cultured B cell hybridomas and T cell lines. Hum Antibodies Hybridomas (in press)

14. Hastie ND, Dempster M, Dunlop MG, Thompson AM, Green DK, Allshire RC (1990) Telomere reduction in human colorectal carcinoma and with aging. Nature 346:866–888

15. James CD, Carlbom E, Dumanski JP, Hansen M, Nordenskjold M, Collins VP, Cavenee WK (1988) Clonal genomic alterations in glioma malignancy stages. Cancer Res 48:5546–5551

16. Jeffreys AJ, Wilson V, Thein SL (1985a) Hypervariable "minisatellite" regions in human DNA. Nature 314:67–73

17. Jeffreys AJ, Wilson V, Thein SL (1985b) Individual-specific "fingerprints" of human DNA. Nature 316:76–79

18. Jenkins RB, Kimmel DW, Moertel CA, Schultz CG, Scheithauer BW, Kelly PJ, Dewald GW (1989) A cytogenetic study of 53 human gliomas. Cancer Genet Cytogenet 39:253–279

19. de Jong D, Voetdijk BMH, Kluin-Nelemans JC, van Ommen GJB, Kluin PM (1988) Somatic changes in B-lymphoproliferative disorders (B-LPD) detected by DNA-fingerprinting. Br J Cancer 58:773–775

20. Kinzler KW, Zehnbauer BA, Brodeur GM, Seeger RC, Trent JM, Meltzer PS, Vogelstein B (1986) Amplification units containing human N-myc and c-myc genes. Proc Natl Acad Sci USA 83:1031–1035

21. Lagoda PJL, Seitz G, Epplen JT, Issinger O-G (1989) Increased detectability of somatic changes in the DNA from human tumours after probing with "synthetic" and "genome-derived" hypervariable multilocus probes. Hum Genet 84:35–40

22. Libermann TA, Nusbaum HR, Razon N, Kris R, Lax I, Soreq H, Whittle N, Waterfield MD, Ullrich A, Schlessinger J (1985) Amplification, enhanced expression and possible rearrangement of EGF receptor gene in primary human brain tumours of glial origin. Nature 313:144–147

22a. Mitreiter R, Epplen C (1992) Self-reactive and antigen specific T cell clones derived from an HLA-DR4$^+$/DR5$^+$ donor: T cell receptors and MHC-restriction patterns. Immunobiol (in press)

23. Nelson M, McClelland M (1989) Effects of site-specific methylation on DNA modification methyltransferases and restriction endonucleases. Nucleic Acids Res 17 [Suppl]:r389–r415

24. Nürnberg P, Epplen JT (1989) "Hidden partials" – a cautionary note. Fingerprint News 1/4:11–12

25. Nürnberg P, Epplen JT (1990) Efficient fingerprint screening of genomic changes in tumors by simple repetitive oligonucleotide probes. Fingerprint News 2/4: 15–17

26. Nürnberg P, Roewer L, Neitzel H, Sperling K, Pöpperl A, Hundrieser J, Pöche H, Epplen C, Zischler H, Epplen JT (1989) DNA fingerprinting with the oligonucleotide probe (CAC)5/(GTG)5: somatic stability and germ line mutations. Hum Genet 84:75–78

27. Nürnberg P, Zischler H, Fuhrmann E, Thiel G, Losanova T, Kinzel D, Nisch G, Witkowski R, Epplen JT (1991) Co-amplification of simple repetitive DNA fingerprint fragments and the EGF receptor gene in human gliomas. Genes Chromosomes Cancer 3:79–88

28. Nürnberg P, Barth I, Fuhrmann E, Lenzner C, Losanova T, Peters C, Pöche H, Thiel G (1991) Monitoring genomic alterations with a panel of oligonucleotide probes specific for various simple repeat motifs. Electrophoresis 12:186–192

28a. Nürnberg P, Thiel G, Weber F, Epplen JT (1992) Changes of telomere lengths in human intracranial tumors. Hum Genet 90 (in press)

29. Pakkala S, Helminen P, Saarinen UM, Alitalo R, Peltonen L (1988) Differences in DNA-fingerprints between remission and relapse in childhood acute lymphoblastic leukemia. Leuk Res 12:757–762

30. Royle NJ, Clarkson RE, Wong Z, Jeffreys AJ (1988) Clustering of hypervariable minisatellites in the proterminal regions of human autosomes. Genomics 3: 352–360

31. Sandberg AA (1990) The chromosomes in human cancer and leukemia, 2nd edn. Elsevier, New York

32. Sambrook J, Fritsch EF, Maniatis T (1989) Molecular cloning: a laboratory handbook. Cold Spring Harbor Laboratory, Cold Spring Harbor

33. Schäfer R, Zischler H, Birsner U, Becker A, Epplen JT (1988) Optimized oligonucleotide probes for DNA fingerprinting. Electrophoresis 9:369–374

34. Smit VTHBM, Cornelisse CJ, de Jong D, Dijkshoorn NJ, Peters AAW, Fleuren GJ (1988) Analysis of tumor heterogeneity in a patient with synchronously occurring female genital tract malignancies by DNA flow cytometry, DNA fingerprinting, and immunohistochemistry. Cancer 62:1146–1152

35. Speth C, Epplen JT, Oberbäumer I (1991) DNA fingerprinting with oligonucleotides can differentiate cell lines derived from the same tumor. In Vitro Cell Dev Biol 27A:646–650

36. Stark GR, Wahl GM (1984) Gene amplification. Annu Rev Biochem 53:447–491

37. Teyssier JR (1989) The chromosomal analysis of human solid tumors. Cancer Genet Cytogenet 37:103–125

38. Thein SL, Jeffreys AJ, Gooi HC, Cotter F, Flint J, O'Connor NTJ, Weatherall DJ, Wainscoat JS (1987) Detection of somatic changes in human cancer DNA by DNA fingerprint analysis. Br J Cancer 55:353–356

39. Wainscoat JS, Thein SL (1985) Polymorphism in human DNA: application to cancer studies. Trends Biochem Sci 10:474–476

40. Wolff RK, Plaetke R, Jeffreys AJ, White R (1989) Unequal crossingover between homologous chromosomes is not the major mechanism involved in the generation of new alleles at VNTR loci. Genomics 5:382–384

41. Wong AJ, Bigner SH, Bigner DD, Kinzler KW, Hamilton SR, Vogelstein B (1987) Increased expression of the epidermal growth factor receptor gene in malignant gliomas is invariably associated with gene amplification. Proc Natl Acad Sci USA 84:6899–6903

42. Zischler H, Nanda I, Schäfer R, Schmid M, Epplen JT (1989) Digoxigenated oligonucleotide probes specific for simple repeats in DNA fingerprinting and hybridization in situ. Hum Genet 82:227–233

Activation of *ras* Oncogenes in Human Tumours

N.R. Lemoine

Molecular Pathology Laboratory, ICRF Oncology Group, Royal Postgraduate
Medical School, Hammersmith Hospital, Du Cane Road, London W12 OHS,
United Kingdom

Introduction

Activated *ras* genes were originally detected using genomic DNA transfection of NIH3T3 cells in focus induction and nude mouse tumorigenesis assays. These assays were technically demanding, expensive and time-consuming. With the realisation that activating point mutations in human tumours are restricted to codons 12, 13 and 61 of the *ras* genes, more rapid techniques were developed for their detection. Since the advent of the polymerase chain reaction DNA amplification technique, the most popular have been the RNase mismatch cleavage assay, oligonucleotide hybridisation, single-strand conformational polymorphism and variations on restriction fragment length polymorphism. We now have the technology to detect *ras* mutations within a single working day, and as I will attempt to illustrate in this chapter, such rapid molecular diagnostics may find application in the practice of modern clinical pathology (Table 1).

The *ras* Genes

In human cells there are three members of the *ras* family, Harvey, Kirsten and N-*ras* (Bos 1989; Lemoine 1990a,b). They encode closely related proteins of 21 kDa that are localised to the inner surface of the plasma membrane and possibly other intracellular membranes by a mechanism that involves complex processing at their C-terminal tails (Hancock et al. 1989, 1990). All *ras* proteins undergo farnesylation of the cysteine residue in the terminal tetrapeptide CAAX (cysteine, two aliphatic amino acids, any amino acid), followed by removal of the last three amino acids and methylation of the cysteine residue. In Harvey and N-*ras* p21 proteins there is then further palmitylation of upstream cysteines to give optimal membrane binding, while in contrast Ki-*ras* p21 relies upon the polylysine

Wagener/Neumann (Eds.)
Molecular Diagnostics of Cancer
© Springer-Verlag Berlin Heidelberg 1993

Table 1. Oncogenes and human cancer

---

1. Understanding basic biology of oncogene action, for instance:
    a) Dominant oncogene products as components of signal transduction pathways
    b) Tumour suppressor gene products as negative transcription regulators and cell cycle control factors
2. Utility of oncogenes and their products as diagnostic markers, for instance:
    a) Ki-*ras* mutations in pancreatic cancer
    b) p53 mutant protein in many human cancers
3. Utility of oncogenes and their products as prognostic markers, for instance:
    a) c-*erb*B-2 amplification/overexpression in breast and ovarian cancers
    b) N-*myc* amplification in neuroblastoma
    c) *ras* mutations in non-small-cell lung cancer
4. Exploitation of oncogenes and their products as therapeutic targets, for instance:
    a) Antisense inhibition of oncogene expression
    b) Mutant EGF receptors in brain tumours as immune targets
    c) Tumour suppressor genes for gene therapy

---

domain that it possesses for full function. Whether these differences in processing result in subtle differences in the connections made with other elements at the membrane is as yet unclear.

The *ras* proteins bind guanine nucleotides, GDP and GTP, and participate in a GTPase cycle (Hall 1990a; Bourne et al. 1990, 1991). There are three conformational states in this cycle: the GDP-bound form of *ras* p21 is inactive but release of the nucleotide converts it to a transient empty state. Under normal intracellular conditions, GTP is more likely to enter the empty guanine nucleotide-binding site since it is present in great excess over GDP, and the protein is induced to adopt its active conformation. This switches back to the inactive GDP-bound state when GTP is hydrolysed. Like most other GTPases, the *ras* proteins have relatively low rate constants for both GDP release and GTP hydrolysis, and full signalling function requires intervention by accessory regulatory proteins. There are guanine nucleotide release or exchange proteins (GNRPs) that catalyse the release of GDP to allow access for GTP, and GTPase-activating proteins (GAPs) that speed up GTP hydrolysis. It is therefore possible to envisage control at two points in the cycle, with more active *ras*-GTP being produced by a reduced rate of GTP hydrolysis or an increased rate of GDP exchange (Macara 1991). An alternative control has been suggested by the discovery that the *nm23* gene product (a nucleotide diphosphate kinase) may be involved in the regulation of *ras*-like proteins and could potentially phosphorylate *ras*-bound GDP to GTP directly (Ruggieri and McCormick 1991).

The *ras* proteins are activated to oncogenic form in human tumours by mutations that produce substitutions at positions 12, 13 or 61. The three-dimensional model of the *ras* protein as solved by Wittinghofer and colleagues (Wittinghofer and Pai 1991) reveals that these three positions lie close to the terminal phosphate bond in GTP, and that the side chains of the amino acids at these positions are critical for the orientation of a water molecule involved in the hydrolysis of GTP. Mutant proteins are no longer able to respond to the upregulating activity of GAPs and remain locked in the active conformation bound to GTP, continuing to transmit the signal to downstream elements of the transduction pathway.

The *ras* Signal Transduction Pathway

The nature of the *ras* signal and the identity of the downstream elements of its signal transduction pathway have recently become rather more clear. Wittinghofer has shown that the *ras* protein changes in conformation during the switch between its inactive and active forms (Wittinghofer and Pai 1991). Two loops are affected by this conformational change: the L4 loop carrying position 61 involved in orientation of the water molecule critical for GTPase activity, and the L2 loop which carries the region known as the effector domain of the *ras* protein between positions 32 and 40. Site-directed mutagenesis to produce substitutions in this effector domain abrogate the transforming action of the oncogenic *ras* and also prevent interactions with the GAP proteins, which is part of the evidence pointing to GAP proteins as effector of *ras* action (Hall 1990b).

The situation has been complicated by the recent discovery that there are two proteins with GTPase-stimulating activity for *ras* p21, both of which are ubiquitously expressed. The first discovered (by McCormick and colleagues) is known as p120 GAP or *ras* GAP and is a cytoplasmic protein of 120 kDa (Adari et al. 1988; Cales et al. 1988). It can increase the rate of GTP hydrolysis by *ras* p21 20 000-fold. The second protein is the product of the gene at the locus responsible for the inherited malignancy neurofibromatosis type 1 (NF1). This gene is a tumour suppressor gene because it is inactivated by deletion or point mutation as part of malignant transformation in this tumour. Expression of a GAP-like domain within the NF1 protein is capable of accelerating the GTPase activity of *ras* p21 like p120 GAP (Ballester et al. 1990; Martin et al. 1990; Xu et al. 1990). Inhibition of the activity of one or both of these GAP proteins will lead to a rise in the amount of p21 *ras* GTP in the cell, while complete loss of their function through genetic inactivation would leave *ras* p21 dependent on its own intrinsic (low) rate of GTP hydrolysis for return to the inactive GDP-bound form.

The most popular of the current models for *ras*-GAP interaction is that proposed by McCormick (Bollag and McCormick 1991). The hypothesis is that p120 GAP and NF1 GAP act both as negative regulators of *ras* and as its effector protein. The evidence for a role as negative regulator is overwhelming – overexpression of GAP suppresses transformation by activated Ha-*ras*, and expression of human GAP in yeast decreases the stimulatory effect of *ras* on adenylyl cyclase. Work by Downward has shown that the amount of active *ras* GTP in T cells stimulated through the CD$_3$ receptor is increased because GAP is inhibited by protein kinase C (Downward et al. 1990). On the other hand, there is evidence that the GAP proteins may be effector molecules for *ras*. For instance, recent observations on the modulation of atrial K$^+$ channels show that addition of *ras* GAP or active *ras* GTP to patches of atrial cell plasma membrane inhibit the coupling of the G$_k$ heterotrimeric G protein to muscarinic receptors and that anti-GAP antibodies block the effect (Yatani et al. 1990). Whatever the precise mechanism involved, this points to GAP having an effector function. It has been suggested that this uncoupling of G proteins from their receptors is a general mechanism of *ras* action.

Interactions with other proteins are also likely (Ullrich and Schlessinger 1990; Cantley et al. 1991). Some membrane receptor tyrosine kinases, such as PDGF and EGF receptors, form signal transduction particles that include p120 GAP and phosphorylate p120 GAP on tyrosine. This phosphorylation event is accompanied by translocation to the membrane from the cytoplasm. Antibody-blocking experiments have shown that the function of the PDGF and EGF receptors in stimulating DNA synthesis requires *ras* p21 action.

In fibroblasts and epithelial cells introduction of oncogenic *ras* p21 rapidly activates protein kinase C, and this is essential for the induction of DNA synthesis. The mechanism for activation of protein kinase C by *ras* is still obscure, although it seems to involve an increase in phosphatidylcholine breakdown from membrane lipids (Marshall 1991; Johnston et al. 1991). However, much of the evidence is contradictory. One rapid response to the upregulation of protein kinase C activity is activation of gene transcription via AP-1 transcription factors consisting of *fos* and *jun* proteins. *ras* stimulates a specific *jun* kinase (Binétruy et al. 1991) and produces a transient spike in the levels of both proteins. Although there does not seem to be a constitutive activation of *fos* and *jun* expression in *ras*-transformed cells, if *fos* expression is blocked by antisense *fos* then the cells revert to a non-transformed phenotype.

There is a second pathway, not involving protein kinase C, that mediates morphological transformation. At the moment it is believed that the c-*raf* and c-*mos* serine/threonine kinases located in the cytoplasm of the cells are involved and that they activate a novel class of transcription factors binding to a *ras*-responsive promoter element with the sequence TGACTCT, while the AP-1 consensus site is TGAGTCA (Owen and Ostrowski 1990).

ras Activation in Tumorigenesis

We are beginning to understand the place of *ras* in signal transduction, but at what point in tumorigenesis is *ras* activation involved? The weight of evidence suggests that it is a relatively early stage in several tumour types. For instance, our own work has shown that at least 75% of cases of carcinoma in situ of the exocrine pancreas harbour activated Ki-*ras* (Lemoine and Hall 1990; Lemoine et al. 1992), and the work of Vogelstein has shown that there is an increasing frequency of mutation during the growth of adenomas of the large bowel before progression to carcinomas (Vogelstein et al. 1988). The work of Padua and others suggests that *ras* mutation is relatively frequent in the premalignant phase of myeloid leukaemia (Padua et al. 1988; Janssen et al. 1987).

So we see that *ras* activation is a critical event in the earliest stages of several major malignancies, but it is by no means a universal finding.

While many tumour types possess mutant *ras* at a frequency approaching 75%, there are others where the frequency is close to zero. There is also a group of tumours with intermediate frequencies (Table 2). The reasons for this striking difference in frequency are not entirely clear, but we believe that it reflects not only the nature of the carcinogens involved in the various tumours, but also how important *ras* proteins are for signal transduction in a particular cell type.

ras Activation and Diagnosis

It will be evident from the preceding paragraphs that *ras* activation by point mutation is a frequent event in several tumour types. For the potential utility of *ras* mutation as a tumour marker it is important that such point mutations do not occur in non-neoplastic conditions, and indeed there are no convincing reports of such events in human tissues (in experimental models in which tumours are induced by exposure to chemical carcinogens *ras* mutations can be demonstrated in the target tissues before the development of tumour cell clones, but this is obviously a different situation). Tumour types which might benefit from improved diagnostic accuracy by the use of *ras* mutation analysis include pancreatic cancer, which is often difficult to differentiate from chronic pancreatitis, particularly in the minute cytological aspirates now popular with radiologists and endoscopists. Bladder cancer

Table 2. *ras* oncogene mutations in human cancers

High frequency	Intermediate	Low frequency
Pancreatic cancer	Melanoma	Breast cancer
Colorectal cancer	Bladder cancer	Female genital tumours
Thyroid cancer	Skin SCC/BCC	Prostate/renal cancers
NSCLC		Gastric cancer
AML		SCLC

NSCLC, non-small cell lung carcinoma; AML, acute myeloid leukaemia; SCC, squamous cell carcinoma; BCC, basal cell carcinoma; SCLC, small cell lung carcinoma.

and leukaemias might be monitored for recurrence by mutation analysis of cell samples at intervals after treatment, and it might also be possible to screen-detect colorectal tumours by such assays. Developments in techniques allied to the polymerase chain reaction (PCR) for DNA amplification hold promise for rapid molecular diagnostics that could be applied in a clinical setting. PCR with mismatched primers to engineer mutation-specific restriction fragment length polymorphisms (RFLP) has been developed for each *ras* oncogene by Kumar and Minna's group, and these allow definite diagnosis of specific *ras* mutations at high sensitivity within a single working day (Kumar and Dunn 1989; Mitsudomi et al. 1991a). The techniques of sequence-specific conformational polymorphism (SSCP) and chemical mismatch cleavage (HOT technique) have some attractions as screening techniques, but the experience of our group and others suggests that both have limitations in this context.

ras Activation and Prognosis

The presence of *ras* mutations can be associated with poor prognosis in some specific tumour types, but does not appear to be a good marker in most human cancers. In non-small-cell lung cancer, particularly adeno-carcinoma, those patients with *ras* mutations at presentation tend to have smaller lesions but a shorter disease-free interval and poorer overall survival than those cases without mutations (Slebos et al. 1990; Mitsudomi et al. 1991b). One report has suggested a possible association of *ras* mutations with poor survival in myelodysplastic syndrome (Yunis et al. 1989), but this has not been confirmed in other studies. It seems that lung cancer may represent a special case in which analysis of *ras* mutation status could give useful prognostic information.

ras-Related Proteins and Phenotypic Reversion

Over the past few years the *ras*-related superfamily has expanded to accommodate 20–30 members identified by protein purification and molecular cloning techniques. The members include R-*ras* (Lowe et al. 1987), *ral* (Chardin and Tavitian 1986), *rab* (Touchot et al. 1987), *rho* (Madaule and Axel 1985), *rac* (Didsbury et al. 1989), and *rap* (Kawata et al. 1988; Kitayama et al. 1989; Pizon et al. 1988a,b) families. Unique GAPs have been identified for *ras* (Adari et al. 1988; Cales et al. 1988; Gibbs et al. 1988), *rho*A (Gibbs et al. 1988), *rap*1 (Kikuchi et al. 1989; Ueda et al. 1989) and *rac* (Diekmann et al. 1991). The *rap*1A and *rap*1B proteins are notable because the putative effector/GAPs interaction domain is identical to that of the *ras* proteins.

The *rap*1A gene was initially cloned as a sequence capable of suppressing transformation by Ki-*ras* in vitro (Kitayama et al. 1989) and more recently shown to inhibit *ras*-induced germinal vesicle breakdown when microinjected into oocytes (Campa et al. 1991). Although the mechanism of suppression is not fully understood, it is known that *rap*1A p21 binds to p120 GAP (which is one of the GAPs for *ras* p21) in vitro in a GTP-dependent manner with very high affinity, without affecting *rap*1A p21 GTPase activity (Frech et al. 1990; Hata et al. 1990; Quillian et al. 1990; Kitayama et al. 1990). If p120 GAP is an effector molecule for *ras* p21, then this would result in inhibition of the transduction of downstream growth signals and suppression of transformation by activated *ras* genes. One tantalising phenomenon is that the expression of *rap*1A can apparently be permanently switched on with concomitant reversion of *ras*-transformed phenotype in vitro by exposure of human epithelial cells to the antibiotic azatyrosine (Kyprianou and Taylor-Papadimitrou, 1992). Whether this is generally applicable strategy will need further intensive study.

Potential Strategies for Reverting *ras* Transformation

The new understanding of the connections of *ras* proteins in signal transduction pathways in normal and transformed cells opens up new avenues for exploration of therapeutic strategies (Table 3). The major interest of our laboratory is in the use of antisense nucleic acids to inhibit expression of oncogenes. Most work has been done on synthetic oligonucleotides, which now show considerable promise as therapeutic agents (Stein and Cohen 1988; Tidd 1991; Uhlmann and Peyman 1990; Calabretta 1991; Dolnick 1991), particularly those with modified linkages that protect against nuclease attack even in vivo (Agrawal et al. 1991). They can be used to specifically inhibit the expression of a mutant *ras* allele (Saisson-Behmoaras et al. 1991; Chang et al. 1991), which obviously has significance for many human tumours, and their major limitation for potential clinical application is the current cost of their manufacture. We are examining the use of retroviral vectors for expression of antisense sequences in a "gene therapy" approach for the treatment of human cancer and developing systems for targeted expression in particular (tumour) tissues (so-called "magic bullet" retroviruses, Stapleford 1991). Certainly such antisense RNA techniques (Takayama and Inouye 1990) can be highly effective in suppressing gene expression and have already been used to revert *ras* transformation of human carcinoma cells in vitro (Mukhopadhyay et al. 1991).

Other levels at which *ras* transformation might be attacked (Table 3) include inhibition of the complex C-terminal processing required for *ras* p21 attachment to membranes elucidated by Hancock and colleagues (Gutierrez et al. 1989; Hancock et al. 1989, 1990; Jackson et al. 1990), and already lovastatin, a cholesterol biosynthesis inhibitor, has been shown to be

Table 3. Potential strategies using mutant *ras* as therapeutic target

1. Interference with synthesis of *ras* p21
   - Antisense oligonucleotides
   - Antisense RNA
2. Interference with processing of *ras* p21
   - Inhibitors of lipid modification
3. Interference with GnP exchange
   - GNRP antagonists
   - GnP analogues
   - Nucleoside diP kinase antagonists?
4. Interference with *ras*-GAP/NF1 interactions
   - Protein kinase C inhibitors
   - Lipid modifiers
   - Gap analogues
5. Competition for GAP/NF1 binding
   - *rap*1A upregulation
6. Exploitation of mutant *ras* peptide/MHCI complexes as immune targets

diP, diphasphate; MHCI, major histocompatibility complex class I.

effective both in vitro (Defeo-Jones et al. 1991) and in vivo (Sebti et al. 1991). This lead compound may be the model for less toxic derivatives that are equally effective in inhibiting *ras* action. Short peptides incorporating the typical CAAX motif of *ras* p21 are also effective in competing for the enzyme system responsible for farnesylation (Reiss et al. 1990), and it might be possible to engineer non-peptide molecules that mimic these peptides for therapeutic use.

Wright and colleagues have explored the potential of GTP derivatives that might interfere with *ras* function (Noonan et al. 1991), but it is not yet possible to predict the utility of this strategy.

Many of the other possibilities listed in Table 3 are speculative, but all are the subject of intense study by many laboratories, academic and industrial, around the world. The prospects for the development of a specific anti-*ras* therapy look fairly encouraging.

References

Adari H, Lowy DR, Willumsen BM, Der CJ, McCormick F (1988) Guanosine triphosphatase activating protein (GAP) interacts with the p21 *ras* effector binding domain. Science 240:518–521

Agrawal S, Temsamani J, Tang JY (1991) Pharmacokinetics, biodistribution, and stability of oligodeoxynucleotide phosphorothioates in mice. Proc Natl Acad Sci USA 88:7595–7599

Ballester R, Marchuk D, Boguski M et al. (1990) The *NF*1 locus encodes a protein functionally related to mammalian GAP and yeast *IRA* proteins. Cell 63:851–859

Binétruy B, Smeal T, Karin M (1991) Ha-*ras* augments c-*jun* activity and stimulates phosphorylation of its activation domain. Nature 351:122–127

Bollag G, McCormick F (1991) Differential regulation of *ras* GAP and neurofibromatosis gene product activities. Nature 351:576–579

Bos JL (1989) *Ras* oncogenes: an overview. Cancer Res 49:4682–4689

Bourne HR, Sanders DA, McCormick F (1990) The GTPase superfamily: a conserved switch for diverse cell functions. Nature 348:125–132

Bourne HR, Sanders DA, McCormick F (1991) The GTPase superfamily: conserved structure and molecular mechanism. Nature 349:117–127

Cales C, Hancock JF, Marshall CJ, Hall A (1988) The cytoplasmic protein GAP is implicated as the target for regulation by the *ras* gene product. Nature 332: 548–551

Calabretta B (1991) Inhibition of protooncogene expression by antisense oligodeoxynucleotides: biological and therapeutic implications. Cancer Res 51:4505–4510

Campa MJ, Chang KG, Molina Y, Vedia L, Reep BR, Lapetine EG (1991) Inhibition of *ras*-induced germinal vesicle breakdown in xenopus oocytes by *rap*-1B. Biochem Biophys Res Commun 174:1–5

Cantley LC, Auger KR, Carpenter C et al. (1991) Oncogenes and signal transduction. Cell 64:281–302

Chang EH, Miller PS, Cushman C et al. (1991) Antisense inhibition of *ras* p21 expression that is sensitive to a point mutation. Biochemistry 30:8283–8286

Chardin P, Tavitian A (1986) The *ral* gene: a new *ras*-related gene isolated by the use of a synthetic probe. EMBO J 5:2703–2708

DeFeo-Jones D, McAvoy EM, Jones RE et al. (1991) Lovastatin selectively inhibits *ras* activation of the 12-0-tetradecanoylphorbol-13-acetate response element in mammalian cells. Mol Cell Biol 11:2307–2310

Didsbury J, Weber RF, Bokoch GM, Evans T, Snyderman R (1989) *Rac*, a novel *ras*-related family of proteins that are botulinum toxin substrates. J Biol Chem 264:16378–16382

Diekmann D, Brill S, Garrett MD et al. (1991) Bcr encodes a gtpase-activating protein for p21 *rac*. Nature 351:400–402

Dolnick BJ (1991) Antisense agents in cancer research and therapeutics. Cancer Invest 9:185–194

Downward J, Graves JD, Warne PH, Rayter S, Cantrell DA (1990) Stimulation of p21^{ras} upon T-cell activation. Nature 346:719–723

Frech M, John J, Pizon V et al. (1990) Inhibition of GTPase activating protein stimulation of *ras*-p21 GTPase by the K*rev*-1 gene product. Science 249:169–171

Gibbs JB, Schaber MD, Alland WJ, Sigal IS, Scolnick EM (1988) Purification of *ras* GTPase activating protein from bovine brain. Proc Natl Acad Sci USA 85:5026–5030

Gutierrez L, Magee AI, Marshall CJ, Hancock JF (1989) Post-translational processing of p21^{ras} is two-step and involves carboxy-methylation and carboxy-terminal proteolysis. EMBO J 8:1093–1098

Hall A (1990a) The cellular functions of small GTP-binding proteins. Science 249:635–640

Hall A (1990b) *Ras* and GAP – who's controlling whom? Cell 61:921–923

Hancock JF, Magee AI, Childs JE, Marshall CJ (1989) All *ras* proteins are polyisoprenylated but only some are palmitoylated. Cell 57:1167–1177

Hancock JF, Paterson H, Marshall CJ (1990) A polybasic domain or palmitoylation is required in addition to the CAAX motif to localise p21^{ras} to the plasma membrane. Cell 63:133–139

Hata Y, Kikuchi A, Sasaki T, Schaber MD, Gibbs JB, Takai Y (1990) Inhibition of the *ras* p21 GTPase-activating protein-stimulated GTPase activity of c-Ha-*ras* p21 by *smg* p21 having the same putative effector domain as *ras* p21s. J Biol Chem 265:7104–7107

Jackson JH, Cochrane CG, Bourne JR, Solski PA, Buss JE, Der CJ (1990) Farnesol modification of Kirsten-*ras* exon 4B protein is essential for transformation. Proc Natl Acad Sci USA 87:3042–3046

Janssen JWG, Steenvoorden ACM, Lyons J et al. (1987) *Ras* gene mutations in acute and chronic myelocytic leukemias, chronic myeloproliferative disorders and myelodysplastic syndromes. Proc Natl Acad Sci USA 84:9228–9232

Johnston C, Morris J, Hall A (1991) The role of *ras* in signal transduction. Biochem Soc Trans 19:296–299

Kawata M, Matsui Y, Kondo J, Hishida T, Teranishi Y, Takai Y (1988) A novel small molecular weight GTP-binding protein with the same putative effector domain as the *ras* proteins in bovine brain membranes: purification, determination of primary structure and characterization. J Biol Chem 263:18965–18971

Kikuchi A, Sasaki T, Araki J, Hata Y, Takai Y (1989) Purification and characterisation from bovine brain cytosol of two GTPase activating protein specific for *smg* p21, a GTP-binding protein having the same effector domain as c-*ras* p21s. J Biol Chem 264:9133–9136

Kitayama H, Sugimoto Y, Matsuzaki T, Ikawa Y, Noda M (1989) A *ras*-related gene with transformation suppressor activity. Cell 56:77–84

Kitayama H, Matsuzaki T, Ikawa Y, Noda M (1990) Genetic analysis of the Kirsten-*ras*-revertant 1 gene: potentiation of its tumour suppressor activity by specific point mutations. Proc Natl Acad Sci USA 87:4284–4288

Kumar R, Dunn LL (1989) Designed diagnostic restriction fragment length polymorphisms for the detection of point mutations in *ras* oncogenes. Oncogene Res 1:235–241

Kyprianou N, Taylor-Papadimitrou J (1992) Isolation of azatyrosine-induced revertants from *ras*-transformed human mammary epithelial cells. Oncogene 7:57–63

Lemoine NR (1990a) *ras* oncogenes in human cancers. In: Sluyser M (ed) The molecular biology of cancer genes. Horwood, Chichester, pp 82–118

Lemoine NR (1990b) The c-*ras* oncogenes and GAP. In: Carney D, Sikora K (eds) Genes and cancer. Wiley, Chichester, pp 19–29

Lemoine NR, Hall PA (1990) Growth factors and oncogenes in pancreatic cancer. Baillières Clin Gastroenterol 4:815–832

Lemoine NR, Jain S, Hughes CM et al. (1992) Ki-*ras* oncogene activation in preinvasive pancreatic cancer. Gastroenterology 102:230–236

Lowe DG, Capin DJ, Delwart E, Sakaguchi AY, Naylor JL, Goddel DU (1987) Structure of the human and murine R-*ras* genes, novel genes closely related to the *ras* proto-oncogenes. Cell 48:137–146

Macara IG (1991) The *ras* superfamily of molecular switches. Cell Sign 3:179–187

Madaule P, Axel R (1985) A novel *ras*-related gene family. Cell 41:31–40

Marshall CJ (1991) How does p21^{ras} transform cells ? Trends Genet 7:91–95

Martin GA, Viskochil D, Bollag G et al. (1990) The GAP-related domain of the neurofibromatosis type 1 gene product interacts with *ras* p21. Cell 63:843–849

Mitsudomi T, Viallet J, Mulshine JL, Linnoila RI, Minna JD, Gazdar AF (1991a) Mutations of *ras* genes distinguish a subset of non-small cell lung cancer cell lines from small-cell lung cancer cell lines. Oncogene 6:1353–1362

Mitsudomi T, Steinberg SM, Oie HK et al. (1991b) *ras* gene mutations in non-small cell lung cancers are associated with shortened survival irrespective of treatment intent. Cancer Res 51:4999–5002

Mukhopadhyay T, Tainsky M, Cavender AC, Roth JA (1991) Specific inhibition of K-*ras* expression and tumorigenicity of lung cancer cells by antisense RNA. Cancer Res 51:1744–1748

Noonan T, Brown N, Dudycz L, Wright G (1991) Interaction of GTP derivatives with cellular and oncogenic *ras* p21 proteins. J Med Chem 34:1302–1307

Owen RD, Ostrowski MC (1990) A nuclear factor that binds to *ras*-responsive enhancer elements is present in human tumour cells. Cell Growth Differ 1:601–606

Padua RA, Carter G, Hughes D et al. (1988) *ras* mutations in myelodysplasia detected by amplification, oligonucleotide hybridisation, and transformation. Leukemia 2:503–510

Pizon V, Chardin P, Lerosey I, Olofsson B, Tavitian A (1988a) Human cDNAs *rap*1 and *rap*2 homologous to the *Drosophila* gene D*ras*3 encode proteins closely related to *ras* in the "effector" domain. Oncogene 3:201–204

Pizon V, Lerosey I, Chardin P, Tavitian A (1988b) Nucleotide sequence of a human cDNA encoding a *ras*-related protein (*rap*1B). Nucleic Acids Res 16:7719

Quillian LA, Der CJ, Clark R et al. (1990) Biochemical characterization of baculovirus-expressed *rap*1A/*Krev*-1 and its regulation by GTPase-activating proteins. Mol Cell Biol 10:2901–2908

Reiss Y, Goldstein JL, Seabra MC, Casey PJ, Brown MS (1990) Inhibition of purified p21^{ras} farnesyl: protein transferase by Cys-AAX tetrapeptides. Cell 62:81–88

Ruggieri R, McCormick F (1991) *ras* and the *awd* couple. Nature 353:390–391

Saisson-Behmoaras T, Tocqué B, Rey I, Chassignol M, Thuong NT, Hélène C (1991) Short modified antisense oligonucleotides directed against Ha-*ras* point mutation induce selective cleavage of the mRNA and inhibit T_{24} cells proliferation. EMBO J 10:1111–1118

Sebti SM, Tkalcevic GT, Jani JP (1991) Lovastatin, a cholesterol biosynthesis inhibitor, inhibits the growth of human H-*ras* oncogene transformed cells in nude mice. Cancer Commun 3:141–147

Slebos RJC, Kibbelaar RE, Dalesio O et al. (1990) K-*ras* oncogene activation as a prognostic marker in adenocarcinoma of the lung. N Engl J Med 323:561–565

Stapleford B (1991) Sexual chemistry. Simon and Schuster, London

Stein CA, Cohen JS (1988) Oligodeoxynucleotides as inhibitors of gene expression: a review. Cancer Res 48:2659–2668

Takayama KM, Inouye M (1990) Antisense RNA. Crit Rev Biochem Mol Biol 25:155–184

Tidd DM (1991) A potential role for antisense oligonucleotide analysis in the development of oncogene targeted cancer chemotherapy. Anticancer Res 10:1169–1182

Touchot N, Chardin P, Tavitian A (1987) Four additional members of the *ras* gene superfamily isolated by an oligonucleotide strategy: molecular cloning of YPT-related cDNAs from a rat brain library. Proc Natl Acad Sci USA 84:8210–8214

Ueda T, Kikuchi A, Ohga N, Yamamoto J, Takai Y (1989) GTPase activating proteins for the *smg*-p21 GTP-binding protein having the same effector domain as the *ras* proteins in human platelets. Biochem Biophys Res Commun 159:1411–1419

Uhlmann E, Peyman A (1990) Antisense oligonucleotides: a new therapeutic principle. Clin Rev 90:544–584

Ullrich A, Schlessinger J (1990) Signal transduction of receptors with tyrosine kinase activity. Cell 61:203–212

Vogelstein B, Fearon ER, Hamilton SR et al. (1988) Genetic alteration during colorectal tumour development. N Engl J Med 319:525–532

Wittinghofer A, Pai EF (1991) The structure of *ras* protein: a model for a universal molecular switch. Trends Biochem Sci 16:382–387

Xu G, Lin B, Tanaka K et al. (1990) The catalytic domain of the neurofibromatosis type 1 gene product stimulates *ras* GTPase and complements *ira* mutants of *S. cerevisiae*. Cell 63:835–841

Yatani A, Okabe K, Polakis P, Haleubeck R, McCormick F, Brown AM (1990) *Ras* p21 and GAP inhibit coupling of muscarinic receptors to atrial K⁺ channels. Cell 61:769–776

Yunis JJ, Boot AJM, Mayer MG, Bos JL (1989) Mechanism of *ras* mutation in myelodysplastic syndrome. Oncogene 4:609–614

Laboratory Techniques in the Investigation of Human Papillomavirus Infection

E.-M. de Villiers

Referenzzentrum für Humanpathogene Papillomviren, Deutsches
Krebsforschungszentrum, Im Neuenheimer Feld 506, W-6900 Heidelberg, FRG

Introduction

Although the existence of a human papillomavirus was demonstrated by
electron microscopy as early as 1949 (Strauss et al. 1949), the plurality of
this group of viruses only became evident about 30 years later (Gissmann
and zur Hausen 1976; Gissmann et al. 1977; Orth et al. 1978). With the
advent of gene technology, methods to clone and characterize the viral
DNA became available, the result of which is a list to date totalling 68
different genotypes of known human papillomaviruses, with probably a
series of yet to be identified types still to come.

Plurality of the Human Papillomaviruses

After the recognition of the existence of different types of human papillomaviruses (HPVs), Orth et al. (1978) demonstrated the presence of different
HPVs in lesions of patients suffering from epidermodysplasia verruciformis.
Here a number of different HPV types can occur in lesions from the same
patient, sometimes even more than one type within one lesion. Table 1
presents a very generalized summary of papillomaviruses detected in skin
lesions (benign and malignant).

   The association of a papillomavirus infection with the development of
genital carcinoma was postulated by zur Hausen (1976). The rapid progress
in papillomavirus research, as well as the development of methods to detect
an HPV infection, have largely been influenced by this association. Not only
have 27 different HPV types been isolated from benign and malignant
genital lesions (Table 2), the various aspects of the hypothesis have been
investigated and verified on a molecular (zur Hausen 1991) and epidemiological (Ley et al. 1991) level. Although the available data can only
serve as an indication (Table 3) (de Villiers 1992), papillomaviruses will

Wagener/Neumann (Eds.)
Molecular Diagnostics of Cancer
© Springer-Verlag Berlin Heidelberg 1993

Table 1. HPV types present in skin lesions

Lesion	HPV type(s)
Verrucae	1, 2, 3, 4, 10, 26, 27, 28, 29, 38, 41, 49, 57, 63, 65
Epidermoid cyst	60
Butcher's wart	2, 7
Epidermodysplasia	2, 3, 10
verruciformis	5, 8, 9, 12, 14, 15, 17, 19, 20, 21, 22, 23, 24, 25, 37, 47, 50
Bowenoid changes	16, 34, 35
Squamous cell	5, 8, 14, 17, 20, 47, 41
carcinoma	

Table 2. HPV types present in genital lesions

Lesion	HPV types
Condylomata acuminata	6, 11, 42, 44, 51, 55, (53), 67
Intraepithelial neoplasia	6, 11, 16, 18, 30, 31, 33, 34, 35, 39, 40, 42, 43, 45, 51, 52, 56, 57, 59, 61, 62, 64
Carcinoma	6, 11, 16, 18, 31, 33, 35, 39, 45, 51, 52, 54, 56, 66

Table 3. HPV types present in tumors of the head and neck

Lesion	HPV types
Papillomatoses	6, 11, 32, 7, 57 (2)
Focal epithelial hyperplasia	13, 32
Carcinoma	2, 6, 11, 16, 18, 30

most probably be shown to play an important etiological role in the development of benign and malignant tumors of the oral mucosa, as well as in head and neck tumors.

The Papillomavirus Genome

The genome of a papillomavirus consists of a number of open reading frames (ORFs), each capable of coding for a messenger RNA which in turn will be translated into a protein. These ORFs are grouped into those coding for proteins active in the nucleus and cytoplasm of the host cell (early proteins), and those coding for viral capsid proteins (late proteins), the L1 and L2. The L1 ORF contains sequences conserved in all papillomaviruses and codes for a group-specific antigen (Jenson et al. 1980;.Kurman et al. 1983). The L2 polypeptide has been suggested to be the type-specific antigen, but such sequences have not yet been identified unequivocally amongst those HPV DNAs sequenced. The E1 has, at least in the bovine papillomavirus, been identified as the gene responsible for the replication and maintenance of the episomal form of the DNA molecule within the host cell (Lusky and Botchan 1984, 1985). The E2 controls the transcription rate of the other early genes (Spalholz et al. 1985; Lambert et al. 1987), such as the E6 and the E7, both of which are known as the genes playing a major role in the malignant transformation of a host cell (Crook et al. 1989). The function of the E4 gene has not been fully understood. The E4 gene product can be detected within the cytoplasmic inclusion granules of papillomatous lesions (Doorbar et al. 1989) and probably acts by binding and disrupting certain cytokeratins (Doorbar et al. 1991). The functions of the other ORFs in the HPV genome have not yet been understood.

The transcription patterns of the ORFs have been studied in great detail (Schwarz et al. 1985; Yee et al. 1985). The localization of the single mRNAs has been demonstrated in genital lesions varying from condyloma acuminata through low-grade and high-grade cervical intraepithelial neoplasia to invasive carcinomas (Stoler et al. 1990). The detection of such transcripts could be of importance in the diagnosis of a lesion, i.e. which grade in the development of malignancy. Another factor of increasing importance in the diagnosis is whether the viral DNA exists as an episomal molecule or whether it is integrated into the cellular genome, in which case the E2 ORF is usually disrupted (Schwarz et al. 1985). The HPV genome is in an episomal state in benign and premalignant genital lesions, but integrated in the majority of invasive carcinomas (Dürst et al. 1985; Matsukura et al. 1989; Cullen et al. 1991).

Definition of an HPV Type

Papillomaviruses need a differentiating cell layer for their circular double-stranded DNA molecule to replicate and their capsid proteins to be synthesized and assembled to form mature virus capsids. Due to these special requirements, papillomaviruses cannot be cultivated in the laboratory on a routine basis. Therefore the methods applied to detect a papillomavirus infection had to be varied from the usual serological detection methods used in other viral infections, to nucleic acid detection. Only recently has it been possible to envisage the use of serological diagnostic methods.

The definition of a papillomavirus type has been based on the degree of DNA sequence homology to other known papillomaviruses (Coggin and zur Hausen 1979). These comparisons have in the past been conducted through hybridization in liquid phase between the DNA genomes of two papillomaviruses. If the homology was higher than 50%, the newly isolated clone was regarded as a subtype of the known papillomavirus. If the homology was lower than 50%, it was classified as a new type. The degree of DNA homology between different HPVs varies considerably. Some types are very closely related, e.g., HPV 6, HPV 11, HPV 13 and HPV 55 and many of the viruses isolated from epidermodysplasia verruciformis lesions, namely HPV 5, HPV 8, HPV 12, HPV 14, HPV 19–23, HPV 25, HPV 47 and HPV 50. Others show almost no degree of homology to any of the known types, e.g., HPV 41. It has become quite evident that the present form of classification is very unsatisfactory. With the increasing number of complete DNA sequences of different HPV types becoming available and more unknown HPV types being isolated and characterized, the possibility of a reclassification of the HPVs is in sight. The present definition of a new type is based on less than 90% sequence homology with any other known HPV type in the ORFs E6, E7 and L1.

Methods of Detecting an HPV Infection

At present, each of the techniques generally used to detect any papillomavirus infection has its advantages and disadvantages. The methods vary greatly in sensitivity and in specificity. None of the tests is the ideal diagnostic tool, but, as long as the question to be answered is well defined, and seen within the limitations of the test used, any of the techniques described below can be used. The inclusion of adequate controls throughout is crucial for the interpretation of results obtained with any test.

Antigen Detection

The papillomavirus group-specific antigen (Jenson et al. 1980) can be detected using commercially available tests. Only highly differentiated cells containing large numbers of viral particles will stain positive, indicating that only lesions in which viral capsids are being produced will be seen as HPV positive. This is misleading insofar as the majority of lesions induced by, for example, HPV 1 will be seen as positive, whereas lesions containing, for instance, HPV 16 will hardly be detected. The different types apparently differ in the frequency with which mature virus particles are being produced. By detecting the group-specific antigen, no distinction can be made as to which type is involved in the infection. The production of individual viral proteins has to be executed with the help of bacterial expression systems. This has delayed progress not only in the development of serological tests, but also in the use of antibodies directed against proteins from individual HPV ORFs to detect infection on histological sections. The situation is due to change in the foreseeable future.

Methods of Hybridization

The methods most commonly used at present all involve the principle of nucleic acid hybridization. Hybridization means, in very simple terms, that two single-stranded fragments of nucleic acid (DNA, RNA or both) will, under the experimental conditions provided, attach to each other like a perfect zipper, if they originate from a mutual parent molecule (high stringency), or imperfectly (i.e., with intermittent loops of discordance), if homology exists only in certain segments of the fragment in question. In the latter case, hybridization conditions can be varied according to the purpose involved. The DNA sequences of HPV 6 and HPV 11 share such a high degree of homology that they will cross-hybridize even under conditions of high stringency. In contrast, HPV 41 DNA will hybridize to any other HPV DNA only if the conditions are such that duplexes will form in spite of numerous mismatches. Such a form of hybridization is, in contrast to high-stringency hybridization, very unstable and can easily be reversed. In order to detect any of these hybridization products, one of the single strands involved ("probe") is either radiolabeled or labeled with a product which can be made visible by additional enzymatic reactions. The nonlabeled single-stranded nucleic acid molecule to be tested is usually fixed to a solid phase.

Probes. To avoid false-positive results, the HPV DNA used as labeled probe should always be separated from the vector used in its production. The latter could hybridize to sequences present in possible bacterial contaminants of the lesion under investigation. The DNA or RNA probe can be labeled using one of the following four methods:

1. Nick translation (Rigby et al. 1977). With the help of DNase, "gaps" are introduced into the double-stranded DNA into which new DNA strands are synthesized with DNA polymerase I incorporating labeled nucleotides. The specific activity of the resulting probe depends on the number and activity of the labeled nucleotides being incorporated.
2. Random-primer synthesis (Feinberg and Vogelstein 1983). A random pool of hexanucleotides being annealed to the DNA molecule act as primers for the enzymatic synthesis of new DNA strands containing the labeled nucleotides. As this can be repeated several times, the resulting DNA probe not only spans the length of the input DNA molecule, but the specific activity of the probe can be higher than obtained when using nick translation.

In both these methods the DNA molecules are melted into single strands before being added to the hybridization solution.

3. Synthesis of single-stranded RNA probes (Melton et al. 1984). HPV DNA, cloned into a plasmid vector carrying bacteriophage RNA poly-

merase binding sites, is transcribed into labeled virus-specific single-stranded RNA molecules after addition of RNA polymerase and the appropriate labeled nucleotides. Such a probe is free of contaminating bacterial sequences and the specific activity obtained is very high.

4. Polymerase chain reaction (PCR) (Mullis and Faloona 1987; Saiki et al. 1988). The PCR can also be used to synthesize labeled probes consisting of short nucleotide sequences. Although the resulting labeling can be extremely high and very specific, oligonucleotides are usually 17–20 nucleotides in length, requiring a modification of regularly used hybridization conditions to assure specific annealing (Sambrook et al. 1989).

New methods for labeling probes, using nonradioactive substances, are constantly being investigated and developed. For example, kits for HPV detection via chemiluminescence will be probably available in the near future.

In Situ Hybridization. Tissue sections are treated to obtain single-stranded DNA molecules. Although the morphology of the tissue can be distinguished after hybridization and the signal can be located within one cell, the sensitivity of this method is rather low. Using a radiolabeled probe, one cell has to contain at least 20–50 DNA genome copies to induce a visible signal upon hybridization (Schneider et al. 1991), whereas the sensitivity decreases to 350 genome copies or more per cell using nonradioactive probes (Beckmann and Myerson 1989; Crum et al. 1988).

The question of whether a distinction can be made between two closely related HPV types present in the same lesion, if the probes are identically labeled, remains unsettled (Herrington et al. 1991). Detection of double or multiple infections will in future be possible with the use of different labeling/detection systems. Although this can at present be done with, for example, digoxigenin- vs biotin-labeled probes, large differences in sensitivity (the latter tenfold less sensitive), as well as undesired background staining, should be taken into account (Morris et al. 1990). A newly recognized advantage of this method is the distinction between integrated molecules (localized signal within the nucleus) and episomal DNA molecules (diffuse hybridization signal over the entire nucleus) (Park et al. 1991; Cooper et al. 1991; Wolber and Clement 1991). Another advantage is the detection of the RNA transcripts of the individual ORFs of an HPV genome. The exact localization of these individual transcripts can help in the diagnosis of the grade of a lesion (Stoler et al. 1990).

Filter In Situ Hybridization. Cells obtained from a scraping or lavage are filtered onto a membrane, denatured in situ and hybridized to a labeled probe (Wagner et al. 1984). Although thousands of samples can be tested with relative ease, this advantage is outweighed by the many disadvantages of this method. These are:

- Very low sensitivity: only 50 HPV genome copies per cell or more can be detected.
- High background combined with nonspecific hybridization, in many cases due to contaminating bacteria or blood and mucus present in the sample.
- The extremely strong dependence of HPV detection on the manner in which the clinical sample was taken (surface area, number and nature of cells, blood and mucus present).
- The limitation of number of HPV types that can be applied as probes.
- Due to resulting background, nonradioactive probes cannot be used.
- Adequate controls are neither available nor can they be imitated in vitro, because the composition of smears differs among individual patients.

Southern Blot Hybridization. For routine diagnostic purposes this method (Southern 1975) is too time-consuming and labor-intensive, although it can be regarded as the test from which the most information can be obtained. Cellular DNA is cut with selected restriction enzyme(s) and electrophoretically separated on an agarose gel. After denaturation, the DNA is transferred and fixed to a membrane. The latter is then hybridized with a labeled HPV probe. The sensitivity ranges between 0.1 and 0.01 HPV genome copies per cell, depending on the amount of cellular DNA and the specific activity of the probe. Questions such as episomal vs integrated, deletions, HPV type involved in the single or even double infections, relatedness of one HPV type to another, detection of unknown HPV types and presence of bacterial infection can all be answered in the minimum number of experiments through the critical choice of hybridization conditions and probes.

Reverse Blot Hybridization. Essentially this method (de Villiers et al. 1986) is similar to the usual Southern blot hybridization, with the difference that the individual HPV types are digested from the vector sequences and these samples then electrophoresed on an agarose gel. After denaturation and transfer to a membrane, hybridization follows using radiolabeled total cellular DNA (1 µg). The number and range of HPV types present on such a blot, i.e., tested for in one experiment, can be varied according to the problem posed. If one of these HPV types is present in the sample tested, a positive signal will be seen in the ca. 8-kb fragment of the specific type. If the HPV type is representative of a group of closely related HPVs, this result could be confirmed with a subsequent Southern blot using the HPV type in question as probe. The advantage of this method is that, with an input of the minimum amount of DNA, only one experiment is needed to test for as large a number of HPV types as wished. The sensitivity is about five HPV genome copies per cell using nylon membrane and a radiolabeled probe. An additional advantage is the detection of contaminating bacterial sequences in the sample – the vector DNA present on the blot will give a positive signal. Only stringent hybridization conditions can be used, due to the fact that the genomes of several of the HPV types cross-hybridize to cellular DNA sequences under relaxed experimental conditions. The specific genome sections/ORFs of the HPVs which share these homologous sequences vary from type to type, rendering a general rule impossible (E.M. de Villiers et al., unpublished results).

Dot/Slot Hybridization. This method differs from the two preceding methods in that the DNA to be fixed onto the membrane is not electrophoretically separated, but fixed as a "dot" (drop) or "slot" (manifold used). Many of the commercial kits available make use of this method. Upon evaluating the results, no estimation can be made of the degree of nonspecific hybridization; therefore, adequate controls should be run parallel and probes should be chosen that definitely do not contain contaminating vector sequences.

Polymerase Chain Reaction

The PCR (Mullis and Faloona 1987; Saiki et al. 1988) is rapidly becoming the most frequently used technique in demonstrating an HPV infection. With the use of selected primers, a certain genome segment is amplified through ca. 30–40 cycles of denaturation/annealing/amplification with a temperature-resistant DNA polymerase. Due to the extremely high sensitivity of this method (one HPV genome copy in 100 000 cells), experimental conditions have to be very tightly controlled to rule out contamination and therefore false-positive results. This contamination could occur at any stage, even as early as the moment the sample is being taken from the patient. Several measures to avoid the possibility of contamination have been introduced, such as the boiling of samples to exclude a step of DNA extraction (Van den Brule et al. 1990). The most generally used primers are chosen in the L1 ORF of the HPV genome (Bauer et al. 1991; Snijders et al. 1990). These primers are seen as consensus primers with which a large spectrum of HPV types can be identified. After the amplification has been done, the resulting sample is separated by gel electrophoresis. Important in avoiding false interpretation of the end product is to separate the amplified DNA by gel electrophoresis with subsequent hybridization to type-specific HPV probes or to digest the amplification samples with restriction enzymes in order to end up with type-specific fragment sizes. Problems can occur if the L1 ORF is deleted in a tumor into which the HPV DNA has been integrated. This is often the case in carcinomas (Schwarz et al. 1985). For this reason, the use of primers in the E6 ORF (Yoshikawa et al. 1991) or E7 ORF (Cornelissen et al. 1989) might be advisable, even though the range of HPV types detectable might be restricted. Irrespective of which primers are being used, a percentage of the probes could still be uninterpretable (Gravitt et al. 1991) or designated as "new HPV types" (Bauer et al. 1991;

Fujinaga et al. 1991). The latter could only be accepted if these "HPV types" could be isolated and characterized as a full-length DNA genome. In view of the short time for which this method has been available, as well as the many influencing factors, for example even the type of fixative used or the fixation time (Greer et al. 1991), it is not surprising that present data the HPV detection rate using the PCR are fairly confusing. Time and experience are needed to sort out the facts. The use of the PCR will probably expand in future; for instance, methods are being developed to amplify the DNA in the intact cells of a histological section (Nuovo, personal communication). By this means the localization of the amplified DNA could be determined within the context of intact cellular morphology.

Comparisons

Many investigators have recently used two or more of the above-mentioned methods on the same samples to compare sensitivity and specificity. As mentioned previously, any test can be used, provided the user is aware of its limitations, positive or negative. A mistake with confusing consequences is the division of clinical samples before comparing different methods. If, for example, a biopsy sample is divided and the separate sections tested individually, completely different results can be obtained. One area of the lesion could contain HPV DNA and the other not. Similarly, if the same instruments are applied to take more than one sample, a carry-over of viral DNA/particles could occur between samples.

The Value of HPV Detection

Several studies have been conducted to determine the prevalence of genital HPV infection in healthy individuals. A negative result obtained from one sample is not necessarily indicative of the absence of an HPV infection. Again, apart from the many varying handling and experimental conditions, a physiological variation can be detected. Multiple sampling of one woman over a period of time reveals fluctuation between positivity and negativity (de Villiers et al. 1992). Although ultimate proof has not yet been obtained, indications are that the HPV infections persist as latent infections, with intermittent cycles of replication or viral production. No drug is yet available to successfully treat or remove these subclinical infections. Therefore, unless a clinical lesion or abnormal cytology has in addition been diagnosed, the detection of an HPV infection should at the present time not be of much consequence. Good evidence exists, especially at the molecular biological (zur Hausen 1991) and epidemiological (Kjaer et al. 1990; Ley et al. 1991) level, to support the notion that an HPV infection alone is not sufficient for the development of a malignant lesion, but that additional host cell modifications are needed for the cascade of events required for malignant conversion. This may be mediated by endogenous as well as by exogenous additional factors.

Acknowledgments. A similar report has been published in the journal *Genitourinary Medicine.* This work was supported in part by the Bundesministerium für Gesundheit, Bonn.

References

Bauer HM, Ting Y, Greer CE, Cambers JC, Tashiro CJ, Chimera J, Reingold A, Manos MM (1991) Genital human papillomavirus infection in female university students as determined by a PCR-based method. JAMA 265:472–477

Beckmann AM, Myerson D (1989) Diagnosis of human papillomavirus and human cytomegalovirus by hybridization histochemistry. In: Tenover FC (ed) DNA probes for infectious diseases. CRC Press, Boca Raton, pp 145–162

Coggin JR, zur Hausen H (1979) Workshop on papillomaviruses and cancer. Cancer Res 39:545–546

Cooper K, Herrington CS, Graham AK, Evans MF, McGee Jo (1991) In situ evidence for HPV 16, 18, 33 integration in cervical squamous cell cancer in Britain and South Africa. J Clin Pathol 44:406–409

Cornelissen MT, van den Tweel JG, Struyk AHB, Jebbink MF, Briet M, van der Noordaa J, ter Schegget J (1989) Localization of human papillomavirus type 16 DNA using the polymerase chain reaction in the cervix uteri of women with cervical intraepithelial neoplasia. J Gen Virol 70:2555–2562

Crook T, Morgenstein JP, Crawford L, Banks L (1989) Continued expression of HPV 16 E7 protein is required for maintenance of the transformed phenotype of cells cotransformed by HPV 16 plus EJ-ras. EMBO J 8:513–519

Crum CP, Nuovo G, Friedman D, Silverstein SV (1988) A comparison of biotin and isotope-labeled ribonucleic acid probes for in situ detection of HPV 16 ribonucleic acid in genital precancers. Lab Invest 58:354

Cullen AP, Reid R, Campion M, Lorincz AT (1991) Analysis of the physical state of different human papillomavirus DNAs in intraepithelial and invasive cervical neoplasm. J Virol 65:606–612

de Villiers E-M (1991) Viruses in cancers of the head and neck. Adv Otorhinolaryngol 46:116–123

de Villiers E-M, Schneider A, Gross G, zur Hausen H (1986) Analysis of benign and malignant urogenital tumors for human papillomavirus infection by labelling cellular DNA. Med Microbiol Immunol (Berl) 174:281–286

de Villiers E-M, Wagner D, Schneider A, Wesch H, Munz F, Miklaw H, zur Hausen H (1992) Human papillomavirus DNA in women without and with cytological abnormalities: results of a five year follow-up study. Gynecol Oncol 44:33–39

Doorbar J, Coneron I, Gallimore PH (1989) Sequence divergence yet conserved physical characteristics among the E4 proteins of cutaneous human papillomaviruses. Virology 172:51–62

Doorbar J, Ely S, Sterling J, McLean C, Crawford L (1991) Specific interaction between HPV-16 E1–E4 and cytokeratins results in collapse of the epithelial cell intermediate filament network. Nature 352:824–827

Dürst M, Kleinheinz A, Hotz M, Gissmann L (1985) The physical state of human papillomavirus type 16 DNA in benign and malignant genital tumors. J Gen Virol 66:1515–1522.

Feinberg FP, Vogelstein B (1983) A technique for radiolabelling DNA restriction endonuclease fragments to high specific activity. Anal Biochem 132:6–13

Fujinaga Y, Shimada M, Okazawa K, Fukushima M, Kato I, Fujinaga K (1991) Simultaneous detection and typing of genital human papillomavirus DNA using polymerase chain reaction. J Gen Virol 72:1039–1044

Gissmann L, zur Hausen H (1976) Human papillomavirus: physical mapping and genetic heterogeneity. Proc Natl Acad Sci USA 73:1310–1313

Gissmann L, Pfister H, zur Hausen H (1977) Human papillomaviruses (HPV): characterization of four different isolates. Virology 76:569–580

Gravitt P, Hakenworth A, Stoerker J (1991) A direct comparison of methods proposed for use in widespread screening of human papillomavirus infections. Mol Cell Probes 5:65–72

Greer CE, Perterson SL, Kiviat NB, Manos MM (1991) PCR amplification from paraffin-embedded tissues. Effects of fixative and fixation time. Am J Clin Pathol 95:117–124

Herrington CS, Graham AK, Flannery DM, Burns J, McGee JO (1991) Discrimination of closely homologous HPV types by nonisotopic in situ hybridization: definition and derivation of tissue melting temperatures. Histochem J 22:545–554

Jenson AB, Rosenthal JD, Olson C, Pass F, Lancaster WD, Shah K (1980) Human papillomavirus: frequency and distribution in plantar and common warts. Lab Invest 47:491–497

Kjaer SK, Engholm G, Teisen C, Haugaard BJ, Lynge E, Christensen RB, Moller KA, Jensen H, Poll P, Vestergaard BF, de Villiers E-M, Jensen OM (1990) Risk factors for cervical human papillomavirus and herpes simplex virus infections in Greenland and Denmark: a population-based study. Am J Epidemiol 131: 669–682

Kurman RJ, Jenson AB, Lancaster WD (1983) Papillomavirus infection of the cervix: II. Relationship to intraepithelial neoplasia based on the presence of specific viral structural proteins. Am J Surg Pathol 7:39–52

Lambert PF, Spalholz BA, Howley PM (1987) A transcriptional repressor encoded by BPV-1 shares common carboxy-terminal domain with the E2 transactivator. Cell 50:69–78

Ley C, Bauer HM, Reingold A, Schiffman MH, Chambers JC, Tashiro CJ, Manos MM (1991) Determinants of genital human papillomavirus infection in young women. J Natl Cancer Inst 83:997–1003

Lusky M, Botchan MR (1984) Characterization of the bovine papillomavirus plasmid maintenance sequences. Cell 36:391–401

Lusky M, Botchan MR (1985) Genetic analysis of bovine papillomavirus type 1 trans-acting replication factors. J Virol 53:955–965

Matsukura T, Koi S, Sugase M (1989) Both episomal and integrated forms of human papillomavirus type 16 are involved in invasive cervical cancers. Virology 172: 63–72

Melton DA, Krieg PA, Rebagliati MR, Maniatis T, Zimm K, Green MR (1984) Efficient in vitro synthesis of biologically active RNA and RNA hybridization probes from plasmids containing a bacteriophage promotor. Nucleic Acid Res 12:7035–7056

Morris RG, Arends MJ, Bishop PE, Sizer K, Duvall E, Bird CC (1990) Sensitivity of digoxigenin and biotin labelled probes for detection of human papillomavirus by in situ hybridization. J Clin Pathol 43:800–805

Mullis KB, Faloona FA (1987) Specific synthesis of DNA in vitro via a polymerase-catalyzed chain reaction. Methods Enzymol 155:335–350

Orth G, Jablonska S, Favre M, Croissant O, Jarzabek-Chorzelska M, Rzesa G (1978) Characterization of two types of human papillomaviruses in lesions of epidermodysplasia verruciformis. Proc Natl Acad Sci USA 87:8170–8174

Park JS, Kurman RJ, Kessis TD, Shah KV (1991) Comparison of peroxidase-labeled DNA probes with radioactive RNA probes for detection of human papillomaviruses by in situ hybridization in paraffin sections. Mod Pathol 4:81–85

Rigby PWJ, Dieckmann M, Rhodes C, Berg P (1977) Labeling deoxyribonucleic acid to high specific activity in vitro by nick translation with DNA polymerase I. J Mol Biol 113:237–251

Saiki RK, Gelfand DH, Stoffel S, Scharf SJ, Higuchi R, Horn GT, Mullis KB, Erlich HA (1988) Primer-directed enzymatic amplification of DNA with a thermostabile DNA polymerase. Science 239:487–491

Sambrook J, Fritsch EF, Maniatis T (1989) Molecular cloning. A laboratory manual. Cold Spring Harbor Laboratory Press

Schneider A, Meinhardt G, Kirchmayr R, Schneider V (1991) Prevalence of human papillomavirus genomes in tissues from the lower genital tract as detected by molecular in situ hybridization. Int J Gynecol Pathol 10:1–14

Schwarz E, Freese UK, Gissmann L, Mayer W, Roggenbuck B, Stremlau A, zur Hausen H (1985) Structure and transcription of human papillomavirus sequences in cervical carcinoma cells. Nature 314:111–114.

Snijders PJF, van den Brule AJC, Schrijnemakers HFJ, Snow G, Meijer CJLM, Walboomers JMM (1990) The use of general primers in the polymerase chain reaction permits the detection of a broad spectrum of human papillomavirus genotypes. J Gen Virol 71:173–181

Southern EM (1975) Detection of specific sequences among DNA fragments separated by gel electrophoresis. J Mol Biol 98:503–517

Spalholz BA, Yang Y-C, Howley PM (1985) Transactivation of bovine papillomavirus transcriptional regulatory element by the E2 gene product. Cell 42:183–191

Stoler MH, Rhodes CR, Whitbeck A, Chow LT, Broker TR (1990) Gene expression of HPV types 16 and 18 in cervical neoplasia. In: Howley PM, Broker TR (eds) Papillomaviruses. Wiley-Liss, New York, pp 1–11

Strauss MJ, Shaw EW, Bunting H, Melnick JL (1949) "Crystalline" virus-like paricles from skin papillomas characterized by intranuclear inclusion bodies. Proc Soc Exp Biol Med 72:46

van den Brule AL, Meijer CJ, Bakels V, Kenemans P, Walboomers J (1990) Rapid detection of human papillomavirus in cervical scrapes by combined general primer-mediated and type-specific polymerase chain reaction. J Clin Microbiol 28:2739–2743

Wagner D, Ikenberg H, Bohm N, Gissmann L (1984) Identification of human papillomavirus in cervical swabs by deoxyribonucleic acid in situ hybridization. Obstet Gynecol 64:767–772

Wolber RA, Clement PB (1991) In situ DNA hybridization of cervical small cell carcinoma and adenocarcinoma using biotin-labeled human papillomavirus probes. Mod Pathol 4:96–100

Yee C, Krishnan-Hewlett Z, Baker CC, Schlegel R, Howley PM (1985) Presence and expression of human papillomavirus sequences in human cervical carcinoma cell lines. Am J Pathol 119:361

Yoshikawa H, Kawana T, Kitagawa K, Mizuno M, Yoshikura H, Iwamoto A (1991) Detection and typing of multiple genital human papillomaviruses by DNA amplification with consensus primers. Jpn J Cancer Res 82:524–531

zur Hausen H (1976) Condylomata acuminata and human genital cancer. Cancer Res 36:794

zur Hausen H (1991) Papillomaviruses-host cell interactions in the pathogenesis of anogenital cancer. In: Brugge J, Curren T, Harlow E, McCormick F (eds) Origins of human cancer II. Cold Spring Harbor Laboratory Press, pp. 685–705.

Detection of Minimal Residual Leukemia by Polymerase Chain Reactions

C.R. Bartram

Abteilung für Kinderheilkunde II, Sektion für Molekularbiologie,
Prittwitzstraße 43, W-7900 Ulm, FRG

Over the past two decades substantial progress has been made in the therapy of human leukemias. This development can be illustrated by the success in the treatment of the most frequent neoplasia in childhood, acute lymphoblastic leukemia (ALL). Application of the BFM protocol induces complete clinical and hematological remission in 98% of patients [15]. However, the reduction of leukemia cell burden below the detection level of morphological examination by no means indicates a complete eradication of the leukemia cell population. Various multicenter trials have empirically established that a 2 years' intensive chemotherapy is necessary in order to achieve a cure rate of, by now, 75% in children. Despite these impressive advances two related clinical problems remain a challenge for today's oncology. On one hand a significant number of the eventually cured patients may in fact receive overtreatment, while on the other hand 25% of the children with ALL still relapse, a clinical course associated with a very poor prognosis.

Thus far the quantity and kinetic behavior of residual leukemic cells have been largely enigmatic due to the general inability of conventional methods, including immunophenotyping, flow cytometry, cytogenetics and Southern blotting to detect neoplastic cells when they form fewer than 1% of the cells in the sample being examined. The use of double-color immunofluorescence has markedly improved the sensitivity with which malignant cells can be detected in leukemias showing phenotypic features that are absent on normal hematopoietic counterparts [2].

More recently the application of polymerase chain reaction (PCR) strategies has opened a new dimension in the analysis of residual disease by permitting the identification of as few as $10^{-4}-10^{-6}$ neoplastic cells [16]. For the first time it has become feasible to assess individual responses to therapeutic efforts at an adequate level of sensitivity. PCR technology may therefore be used to predict impending relapse prior to clinical manifestation, thereby permitting initiation of alternative treatments with as small

Wagener/Neumann (Eds.)
Molecular Diagnostics of Cancer
© Springer-Verlag Berlin Heidelberg 1993

leukemia burden as possible, but also to avoid excessive toxicity in patients who may already be cured. In the following I will briefly summarize our experience with PCR methods in monitoring therapeutic effectiveness in leukemia patients. Preliminary as they are, these data point to a pivotal role of PCR approaches in defining a rationale for the development of case-adapted treatment modifications.

Clonospecific TCRδ Probes from ALL Patients

Leukemia cells of ALL patients are characterized by a unique pattern of immunoglobulin (Ig) and/or T-cell receptor (TCR) gene recombinations. Based on the individual immunogenotype different PCR strategies have recently been proposed for the evaluation of minimal residual leukemia [4, 6, 17, 18]. A method initiated in our laboratory proceeds from the observation that the majority of ALL patients exhibit TCRδ rearrangements and/or deletions [6, 20]. Moreover, analyses of more than 500 patients enrolled in the German Multicenter ALL Trials for Children (BFM) and Adults (BMFT) established a preferential recombination pattern depending on the immunophenotype, with a prevalence of $V\delta_1 DJ\delta_1$ (29%), $V\delta_2 DJ\delta_1$ (11%) and $D\delta_2 J\delta_1$ (19%) rearrangements in T-ALL, in contrast to a predominance of $V\delta_2 D\delta_3$ (52%) and $D\delta_2 D\delta_3$ (16%) recombinations in B-precursor leukemias. Based on the restricted pattern of TCRδ recombinations, on one hand, and the enormous junctional diversity due to insertion of N-region nucleotides, on the other hand, we have amplified and isolated TCRδ junctional regions of ALL patients and used them as clonospecific probes.

Thus far we have generated probes from 63 ALL patients (52 children, 11 adults) including 40 common ALL and 23 T-ALL cases. The detection limit of each probe as determined by at least two independent dilution and amplification series varied, but detection of 10^{-4}–10^{-6} neoplastic cells was possible in 56 cases. Probes derived from the other seven patients exhibited a detection level of about 10^{-3}. In any event, the level of sensitivity still exceeds that of conventional methods by at least an order of magnitude. Consecutively we used the clonospecific probes to analyze bone marrow (BM) and peripheral blood (PB) samples obtained from the patients during complete clinical remission. Since in all instances where both BM and PB specimens from a patient were available the latter contained fewer residual cells, if any, I will only refer to data regarding BM samples.

In a first series of 47 patients (Table 1) clonospecific probes identified residual leukemia cells in the majority of cases during consolidation therapy. A significant number of patients still showed minimal residual disease at a level of 10^{-3}–10^{-6} cells during the phase of maintenance therapy, while patients generally lacked evidence of leukemia cells after termination of treatment. The exception was a patient who relapsed clinically 8 months after the PCR analysis.

Table 1. PCR analysis of ALL patients in complete clinical remission using clono-specific TCRδ probes

Treatment phase	Months after diagnosis	No. of samples[a]	PCR status	
			Positive	Negative
Consolidation	1–6	14	10	4
Maintenance	7–24	30	9	21
Termination	>24	32	1	31

[a] Evaluation of 76 bone marrow samples obtained from 47 patients.

However, the data obtained from single PCR analyses have little clinical relevance. Much more important appears the knowledge of the dynamic behavior of leukemia cells in individual patients as determined through serial PCR analyses. In fact, longitudinal studies have disclosed marked individual differences in the intervals between achievement of clinical remission and eradication of residual leukemia below the detection level of PCR (Fig. 1). Interestingly, these disparities in the reduction of neoplastic cells do not correlate with known risk factors, but rather decipher a novel component of the individual response to chemotherapy. While a steady, albeit prolonged reduction of leukemia cells may be associated with a favorable prognosis, a continuous increase of blasts predicts disease recurrence (Fig. 1). Similar results have been obtained by other investigators [10, 14, 19].

We also analyzed a second group of 16 ALL patients with leukemia relapse for whom BM samples were available from the time of complete remission. In 12 cases persistence and consecutive increase of leukemia cells could be identified 6–12 months prior to clinical manifestation (Fig. 1). In three patients, multiple specimens obtained during maintenance therapy first revealed negative PCR results but then scored positive 3–6 weeks before relapse. One might speculate that the focal nature of residual disease interfered with an earlier detection of relapsing blasts in these cases [11]. In the one remaining patient PCR analysis failed to detect leukemia relapse due to a continuing recombination at the TCRδ locus representing the clonospecific probe. This technical pitfall highlights a limitation of any PCR strategy based on the immunogenotype of leukemia cells.

The data summarized thus far demonstrate that clonospecific TCRδ probes in conjunction with PCR analysis constitute a reliable tool to address the problem of minimal residual leukemia. However, it should be emphasized that a combination of different approaches will be required to analyze most, if not all, ALL patients and to confirm data derived from an individual method. One example is the concurrent analysis of remission samples by double-color immunofluorescence and PCR technology. Investigation of seven suitable cases by clonospecific TCRδ probes confirmed the immunological evidence for minimal residual disease in all instances [3]. On the

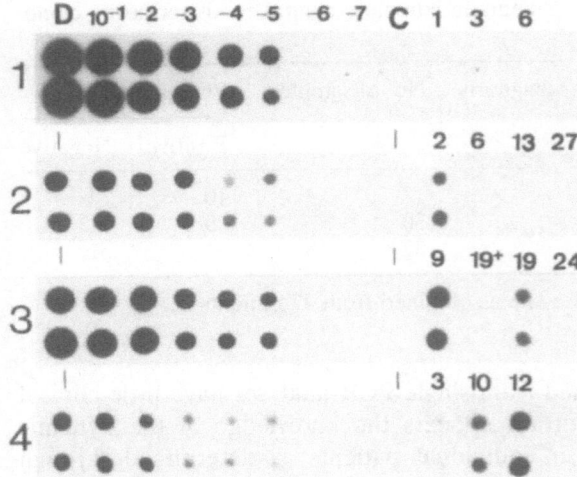

Fig. 1. Detection of minimal residual disease in four ALL patients using clonospecific TCRδ probes. *Left*: DNA of leukemia cells at diagnosis (*D*) was diluted into peripheral blood cell DNA of four healthy individuals (*C*) at 10^{-1} to 10^{-7}, establishing a detection limit of approximately 10^{-5} leukemia cells in all cases. *Right*: Bone marrow DNA samples obtained during the patients' complete clinical/hematological remission were also analyzed (numbers indicate months after diagnosis). After amplification, 20-ng DNA fractions were spotted and hybridized to the clonospecific probes. Note marked differences among the patients in kinetic behavior of residual leukemia cell populations. The month 12 sample of case 4 was obtained at the time of clinical relapse

other hand, PCR techniques may be slightly more sensitive. Accordingly, some bone remission samples scored positive at a $10^{-5}/10^{-6}$ level by PCR analysis only. Another possibility is the combination of different PCR strategies based on distinct genetic markers. In this context a considerable number of chromosomal defects characterizing hematopoietic neoplasias have recently been defined at the molecular level and thereby become accessible to PCR analysis.

BCR-ABL Rearrangement in ALL

The Philadelphia (Ph) translocation was originally discovered in chronic myelocytic leukemia (CML), but it is also the most frequent chromosomal abnormality in adult ALL. Cytogenetically the Ph chromosomes in ALL and CML are indistinguishable, but on the molecular level two distinct subtypes have been defined. The breakpoints of CML patients have been mapped almost exclusively to the major breakpoint cluster region (M-bcr) on chromosome 22, while the majority of Ph-positive ALL patients show a translocation of ABL sequences into the minor breakpoint cluster region (m-bcr) of the BCR gene [9]. Based on PCR analyses of more than 300 ALL patients we recently found a remarkably high incidence of 55% of BCR-ABL-positive cases among adult patients, in contrast to only 6% of children with primary ALL [12]. This study also comfirmed the very poor prognosis associated with this leukemia subtype and provided a rationale for the substantially poorer outcome of ALL in adults than in children.

We have studied the BCR-ABL status in 12 ALL patients during complete clinical remission 3–14 months after initiating chemotherapy according to the BMFT protocol. Interestingly, BM samples of three patients were PCR-negative (Fig. 2). This result suggests that even standard treatment regimens can reduce minimal residual disease to below the detection level of PCR in at least some of these high-risk patients. This preliminary observation merits further evaluation.

Moreover, BM samples of nine ALL patients were studied concurrently by application of clonspecific TCRδ probes and amplification of BCR-ABL

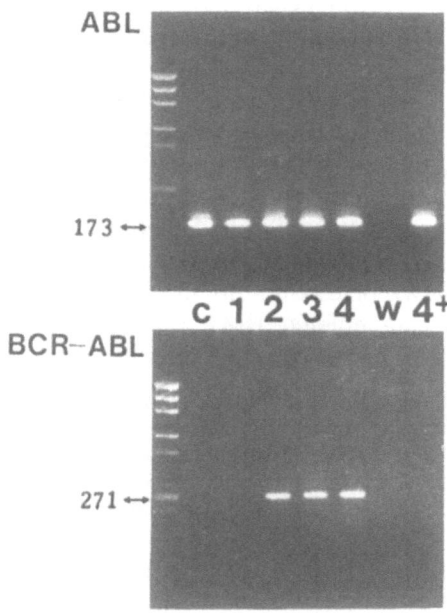

Fig. 2. PCR analysis of four Ph-positive ALL patients with a m-bcr breakpoint during complete clinical remission. A healthy individual (c) and a water sample (w) were included as controls. Upon generation of cDNA from RNA samples, two aliquots were amplified by oligomers detecting either normal ABL fragments (173 bp) or chimeric BCR-ABL products (271 bp). Residual leukemia cells are observed in the bone marrow of patients 2, 3 and 4 but not in that of patient 1. The patients had been in remission for 4, 7, 2 and 8 months respectively. Patient 4 became PCR-negative 4 months after the initial PCR analysis (*lane 4+*)

sequences. Mutual confirmation of the data derived from both approaches was obtained for all samples tested.

CML After Bone Marrow Transplantation

It appears unlikely that the lessons that will be learned by analyzing minimal residual disease in a particular leukemia entity can be generalized. Rather, this issue awaits independent evaluation for each hematopoietic neoplasia, taking into account the biological characteristics of the malignancy under investigation as well as specific treatment modalities. This view is supported by recent data on minimal residual leukemia in CML patients after allogeneic bone marrow transplantation (BMT) [8].

Cytogenetic and Southern blot analyses had previously demonstrated that the persistence and/or fluctuation of Ph-positive cells over several years after transplantation may not necessarily be associated with a clinical relapse [1]. We recently evaluated the remission status of CML patients for whom (a) multiple PCR analyses had been performed longer than 6 months after BMT, (b) clinical follow-up was known for more than 12 months after initial PCR analysis and (c) cytogenetic and Southern blot analysis indicated a complete remission.

The results are summarized in Table 2. All patients who had received T-cell-depleted marrow for prophylaxis of graft-versus-host disease (GVHD) exhibited residual disease years after transplantation. Four of these patients experienced cytogenetic or clinical relapses during follow-up. This result suggests that T-cell depletion interferes with the complete eradication of CML cells, probably due to the lack of graft-versus-leukemia effect. This view corresponds to the well-established increase in clinical relapse after T-cell-depleted BMT [5]. Along the same lines, GVHD may influence

Table 2. PCR analysis of Ph-positive CML patients after bone marrow transplantation

	n	Clinical characteristics	PCR status	Follow-up
I	7	T-cell-depleted BMT	7 positive (56–84 months after BMT)	2 cytogenetic relapses 2 clinical relapses
II	29	Unmanipulated BMT		
a	6	Long-term survivors (>5 years)	6 negative	6 CCR
b	23	Initial PCR analysis 7–37 months after BMT	10 positive	2 PCR-negative 2 cytogenetic relapses 3 clinical relapses
			13 negative	1 PCR-positive 1 cytogenetic relapse

BMT, bone marrow transplantation; CCR, complete clinical remission.

Table 3. Influence of graft-versus-host disease (GVHD) on the eradication of minimal residual disease in CML patients after bone marrow transplantation

n	GVHD	PCR status	Follow-up
4	Absent	4 positive	1 cytogenetic relapse 1 clinical relapse
15	Present	3 positive	1 cytogenetic relapse 1 clinical relapse
		12 negative	12 CCR

CCR, complete clinical remission.

the elimination of minimal residual disease in CML patients receiving an unmanipulated marrow (Table 3). Thus, in the absence of GVHD patients remain PCR-positive, while the majority of patients suffering from GVHD lack evidence of residual leukemia.

Among the 29 patients who had received an unmanipulated transplant, all patients in complete remission for longer than 5 years showed no residual disease (Table 2). This observation confirms the findings of a previous study [13] and indicates that complete elimination of the malignant cell population is feasible and may constitute a prerequisite for curing CML. Moreover, a PCR-negative result beyond month 6 after BMT seems to be associated with a more favorable prognosis than a BCR-ABL-positive status this time (Table 2) [7].

The data discussed above highlight the relevance of immunological mechanisms in the eradication of residual leukemia and also indicate substantial differences in the biology of minimal residual disease in this group of patients as compared to ALL patients receiving polychemotherapy.

Prospects

Within the near future a broad spectrum of PCR methods based on distinct molecular markers will allow the analysis of minimal residual disease in the majority of leukemia patients. Additional methods such as double-color immunofluorescence or interphase in situ hybridization may be used as complementary approaches. Since all techniques bear relevant limitations and specific advantages the combined usage of different strategies will be required to evaluate this issue and to balance interpretations derived from individual methods. It will also be necessary to standardize the various techniques and to control carefully for false-positive or false-negative results.

At the present time the clinical significance of detecting as few as 10^{-6} residual neoplastic cells is far from being settled. The answer can only come from prospective analyses enrolling large numbers of patients. The tools to tackle the problem of minimal residual disease are available. Since the data derived from such analyses may have a profound impact on the treatment of hematopoietic neoplasias, including the development of individualized protocols, it appears appropriate to initiate these studies without further delay.

Acknowledgments. I thank my coworkers T.E. Hansen-Hagge, J.W.G. Janssen, M. Schmidtberger, C. Tell and S. Yokota, who performed the molecular genetic analyses summarized above. I am grateful to A. Biondi, D. Campana and R. Arnold, as well as the participants in the Multicenter ALL Trials for Children (BFM) and Adults (BMFT), namely D. Hoelzer, W.D. Ludwig, H. Riehm and E. Thiel, for fruitful cooperation. This work was supported by grants from the Deutsche Forschungsgemeinschaft, the Deutsche Krebshilfe, and the Förderkreis für tumor- und leukämiekranke Kinder Ulm.

References

1. Arthur CK, Apperley JF, Guo AP, Rassool F, Gao LM, Goldman JM (1988) Cytogenetic events after bone marrow transplantation for chronic myelogenous leukemia in chronic phase. Blood 71:1179–1186
2. Campana D, Coustan-Smith E, Janossy G (1990) The immunologic detection of residual disease in acute leukemia. Blood 76:163–171
3. Campana D, Yokota S, Coustan-Smith E, Hansen-Hagge TE, Janossy G, Bartram CR (1990) The detection of residual acute lymphoblastic leukemia cells with immunologic methods and polymerase chain reaction: a comparative study. Leukemia 4:609–614
4. D'Auriol L, MacIntyre E, Galibert E, Sigaux (1989) In vitro amplification of T cell γ gene rearrangements: a new tool for the assessment of minimal residual disease in acute lymphoblastic leukemias. Leukemia 3:155–158
5. Goldman JM, Gale RP, Horowitz MM, Biggs JC, Champlin RE, Gluckman E, Hoffman RG, Jacobsen SJ, Marmont AM, McGlave PB, Messner HA, Rimmt A, Rozman C, Speck B, Tura S, Weiner RS, Bortin MM (1988) Bone marrow transplantation for chronic myelogenous leukemia in chronic phase: increased risk of relapse associated with T cell depletion. Ann Intern Med 108:806–814
6. Hansen-Hagge TE, Yokota S, Bartram CR (1989) Detection of minimal residual disease in acute lymphoblastic leukemia by in vitro amplification of rearranged T-cell receptor δ chain sequences. Blood 74:1762–1767
7. Hughes TP, Morgan GJ, Martiat P, Goldman JM (1991) Detection of residual leukemia after bone marrow transplant for chronic myeloid leukemia: role of polymerase chain reaction in predicting relapse. Blood 77:874–878
8. Hughes TP, Ambrosetti A, Barbu V, Bartram CR, Battista R, Biondi A, Chiamenti A, Cimino G, Ernst P, Frassoni F, Gasparini P, Gentlilni J, Gluckman E, Grosveld G, Guerrasio A, Hegewich S, Janssen JWG, Keating A, Lo Coco F, Martiat P, Martinelli G, Mills G, Morgan G, Nadali G, Pelicci PG, Perona G, Pignatti PF, Richard P, Saglio G, Trabetti E, Turco A, Veneri D, Zaccaria A, Zander A, Goldman JM (1991) Clinical value of PCR in diagnosis and follow-up of leukemia and lymphoma: report of the 3rd workshop of the Molecular Biology/BMT Study Group. Leukemia 5:448–451

9. Kurzrock R, Gutterman JU, Talpaz M (1988) The molecular genetics of Philadelphia chromosome-positive leukemias. N Engl J Med 319:990–998
10. MacIntyre EA, d'Auriol L, Duparc C, Leverger G, Galibert F, Sigaux F (1990) Use of oligonucleotide probes directed against T cell antigen receptor gamma delta variable-(diversity)-joining junctional sequences as a general method for detecting minimal residual disease in acute lymphoblastic leukemias. J Clin Invest 86:2125–2135
11. Martens ACM, Schultz FW, Hagenbeek A (1987) Nonhomogenous distribution of leukemia in the bone marrow during minimal residual disease. Blood 70:1073–1978
12. Maurer J, Janssen JWG, Thiel E, van Denderen J, Ludwig WD, Aydemir Ü, Heinze B, Fonatsch C, Harbott J, Reiter A, Riehm H, Hoelzer D, Bartram CR (1991) Detection of chimeric BCR-ABL genes in acute lymphoblastic leukaemia by polymerase chain reaction. Lancet 337:1055–1058
13. Morgan GJ, Hughes T, Janssen JWG, Gow J, Guo AP, Goldman JM, Wiedemann LM, Bartram CR (1989) Polymerase chain reaction for detection of residual leukaemia. Lancet 1:928–929
14. Neale GAM, Menarguez J, Kitchingman GR, Fitzgerald TJ, Koehler M, Mirro J, Goorha RM (1991) Detection of minimal residual disease in T-cell acute lymphoblastic leukemia using polymerase chain reaction predicts impending relapse. Blood 78:739–747
15. Riehm H, Gadner H, Henze G, Kornhuber B, Lampert F, Niethammer D, Reiter A, Schellong G (1990) Results and significance of six randomized trials in four consecutive ALL-BFM studies. In: Büchner T, Schellong G, Hiddemann W, Ritter J (eds) Acute leukemias: II. Prognostic factors and treatment strategies. Springer, Berlin Heidelberg New York, pp 439–450
16. Saiki RK, Gelfand DH, Stoffel S, Scharf SJ, Higuchi R, Horn GT, Mullis KB, Erlich HA (1988) Primer-directed enzymatic amplification of DNA with a thermostable DNA polymerase. Science 239:487–491
17. Veelken H, Tycko B, Sklar J (1991) Sensitive detection of clonal antigen receptor gene rearrangements for the diagnosis and monitoring of lymphoid neoplasms by a polymerase chain reaction-mediated ribonucleose protection assay. Blood 78:1318–1326
18. Yamada M, Hudson S, Tournay O, Bittenbender S, Shane SS, Lange B, Tsujimoto Y, Caton AJ, Rovera G (1989) Detection of minimal residual disease in hematopoietic malignancies of the B-cell lineage by using third-complementary-determining region (CDR-III) specific probes. Proc Natl Acad Sci USA 86:5123–5127
19. Yamada M, Wasserman R, Lange B, Reichard BA, Womer RB, Rovera G (1990) Minimal residual disease in childhood B-lineage lymphoblastic leukemia. Persistence of leukemic cells during the first 18 months of treatment. N Engl J Med 323:488–495
20. Yokota S, Hansen-Hagge TE, Ludwig WD, Reiter A, Raghavachar A, Kleihauer E, Bartram CR (1991) The use of polymerase chain reactions to monitor minimal residual disease in acute lymphoblastic leukemia patients. Blood 77:331–339

A Dominant Metastogene Confers Metastatic Potential to Tumor Cells

H. Ponta,[1] M. Zöller,[2] K.-H. Heider,[1] S. Seiter,[2] W. Rudy,[1]
M. Hofmann,[1] C. Tölg,[1] and P. Herrlich[1]

1 Kernforschungszentrum Karlsruhe, Institut für Genetik und Toxikologie PO Box 3640, W-7500 Karlsruhe 1, FRG
2 Institut für Radiologie und Pathophysiologie, Deutsches Krebsforschungszentrum, Im Neuenheimer Feld 280, W-6900 Heidelberg 1, FRG

Cancer's ability to form metastases accounts for the majority of cancer deaths. Although many years of cancer research have generated an impressive list of properties detected on aggressive metastatic cancer cells, the order of events, the causal interrelationships and the decisive contributions that are limiting in the generation of the metastatic phenotype remain ill understood. One reason for this state of ignorance may be the tremendous complexity of the metastatic process, even exceeding that of carcinogenic transformation.

Putative Functional Requirements for the Metastatic Process

A minority of human cancer cells (such as basaliomas) rarely metastasize. In other tumors, stages with different metastatic potential can be distinguished. For instance, breast cancer of less than 1 cm in diameter has usually not yet spread. In colon cancer, particularly in the high-cancerincidence syndrome polyposis coli, various transitions from benign polyps to noninvasive locally growing forms to invasive and metastatic cancer have been described. Thus it is reasonable to propose tumor progression to be a process separate from cancerous transformation. What properties, then, do tumor cells need to acquire, in addition to transformation, to be able to metastasize? The following listing represents a rather hypothetical attempt to accompany cells through the process.

Pathologists classify tumors according to features they share with normal cells, that is the cells the tumor has probably originated from. We now know that differentiated cells can still be transformed if an oncogene is expressed under the control of a differentiation-specific promoter (e.g., as a transgene: Steward et al. 1984; Hanahan 1985; Sinn et al. 1987; Andres et al. 1987; Schoenenberger et al. 1988; Bailleul et al. 1990). A localized tumor often maintains the cell-cell and cell-matrix interaction that normal tissue would

Wagener/Neumann (Eds.)
Molecular Diagnostics of Cancer
© Springer-Verlag Berlin Heidelberg 1993

also exhibit. Initiation of the metastatic spread will therefore involve "deadhesion", loss of these contacts. Examples may be loss of the integrin VLA-5 ($\beta 1\alpha 5$), which mediates contact to fibronectin, and of E cadherin, which is engaged in homotypic interactions between cells (Giancotti and Ruoslathi 1990; Behrens et al. 1989).

In the subsequent phase of migration, both enzymatic properties and new and perhaps transient contacts (adhesions) will be necessary. In addition, directed migrating behavior may involve perception of a chemoattractant gradient. Balanced matrix degradation by surface-bound proteases will permit cells to leave the encapsulated tumor area and pass through stromal matrix. Movement along matrix structures will bring the cells close to the major barrier: the basal membrane. It is still an open question whether lymphatic vessels are sufficiently fenestrated to permit easy access (Hartveit 1990). It is safe to postulate that metastatic cells bind specifically to the basal membrane, e.g., by collagen receptors (VLA-2 and VLA-3), and expose the enzymes needed for destruction. Many tumor cells in fact express elevated levels of VLA-2 or -3 (Klein et al. 1991; Plantefaber et al. 1989), and VLA-2 by transfection increased the potential to metastasize of rhabdomyosarcoma cells in a xenograft transplantation model (Chan et al. 1991). Several enzymes, e.g., plasminogen activator, stromelysin and collagenase (type I, type IV), have been assayed and found to promote the invasive behavior of tumor cells (Ossowski and Reich 1983; Dano et al. 1985; Matrisian et al. 1986; Tryggvason et al. 1987).

Most carcinomas of man or rat spread lymphogenically with preference. How survival and passage through the lymphatic system is made possible remains unknown. After multiplication in draining lymph nodes, cells reach the bloodstream. Somewhere in this passage it seems advantageous to change the MHC haplotype expressed (Wallich et al. 1985; Plaksin et al. 1988; Bernards et al. 1983; Schrier et al. 1983). Microaggregation by cell-cell adhesion or via lectins may promote survival (Raz and Lotan 1987).

It is believed that metastatic cells adhere to endothelial cells. Adherence would fulfill a precondition for extravasation out of the bloodstream. Since different tumors possess a clear preference for organs they metastasize to, one could argue that they distinguish organs on the level of endothelial recognition (review: Pauli et al. 1990). A precedent would be the physiological process of lymphocyte homing (Stoolman 1988; Kieran et al. 1989; Duijvestijn and Hamann 1989). Lymphocytes are thought to return to the lymphoid organs they came from via specific interaction with endothelium. Alternatively, tissue-specific outgrowth of metastases may be regulated after arrival of the cells in their target tissues (reviewed in Nicolson 1988). After all, outgrowth in a new tissue is a demanding process, involving, for instance, the correct set of growth factor receptors and perhaps even autocrine secretion of growth factors.

Isolation of a Metastases-Specific cDNA

It is clear, then, that the metastatic process is complex and involves many molecular properties. Initially, we speculated that there ought to be regulatory genes that control such properties. Attempts at transferring a regulatory gene by genomic and cDNA transfection failed. We therefore set out to identify proteins which are differentially expressed in metastasizing cells and which might trigger just one of the various steps required for metastasis formation.

Monoclonal antibodies were raised against surface proteins of a metastatic and a nonmetastatic variant of a pancreatic carcinoma of the rat (Matzku et al. 1983), and those antibodies were selected that reacted with either the metastatic or the nonmetastatic variant (Matzku et al. 1989; Günthert et al. 1991). The monoclonal antibody to be described here, 1.1ASML, stains the surface of the highly metastatic cell line BSp73 ASML. The epitope is carried by four defined, highly glycosylated proteins of 120–200 kDa apparent molecular weight and is not detectable on the nonmetastatic variants of the same tumor (Günthert et al. 1991). The epitope turned out to consist of primary amino acid sequence, which permitted direct bacterial expression-cloning of cDNA.

The sequence of the complete cDNA clone, obtained after several efforts, revealed various types of information on this metastases-associated molecule. It codes for one of the four proteins detected in the cellular membrane of ASML cells. By screening through recent issues of the journals *Cell* and *Proceedings of the National Academy of Sciences of the USA*, a homologous sequence was found: the sequence of the lymphocyte glyco-protein CD44 (Stamenkovic et al. 1989; Goldstein et al. 1989; Idzerda et al. 1989; Nottenburg et al. 1989). Homology existed in the 5' portion of the sequence and in the 3' portion that coded for a transmembrane segment of the protein and a cytoplasmic tail. The middle portion encoding the epitope was, however, not represented in CD44 (Günthert et al. 1991). Apparently the metastatic cells expressed a longer splice variant of CD44. By amplifying cDNA of several metastatic rat and mouse tumors we found a whole range of different splice variants (Fig. 1). To date the count of putative exons that occurred as extra sequences inserted in the extracellular portion of lymphocyte CD44 has reached 10. Thus metastasizing tumors appear to draw on a sizable exon pool probably existing for defined physiologic functions.

The Variant of CD44 Acts as a Dominant Metastogene

To examine whether expression of the variants of CD44 were coincidental or causally related to the metastatic phenotype, we isolated several trans-fectants of the nonmetastatic pancreatic carcinoma line using the variant

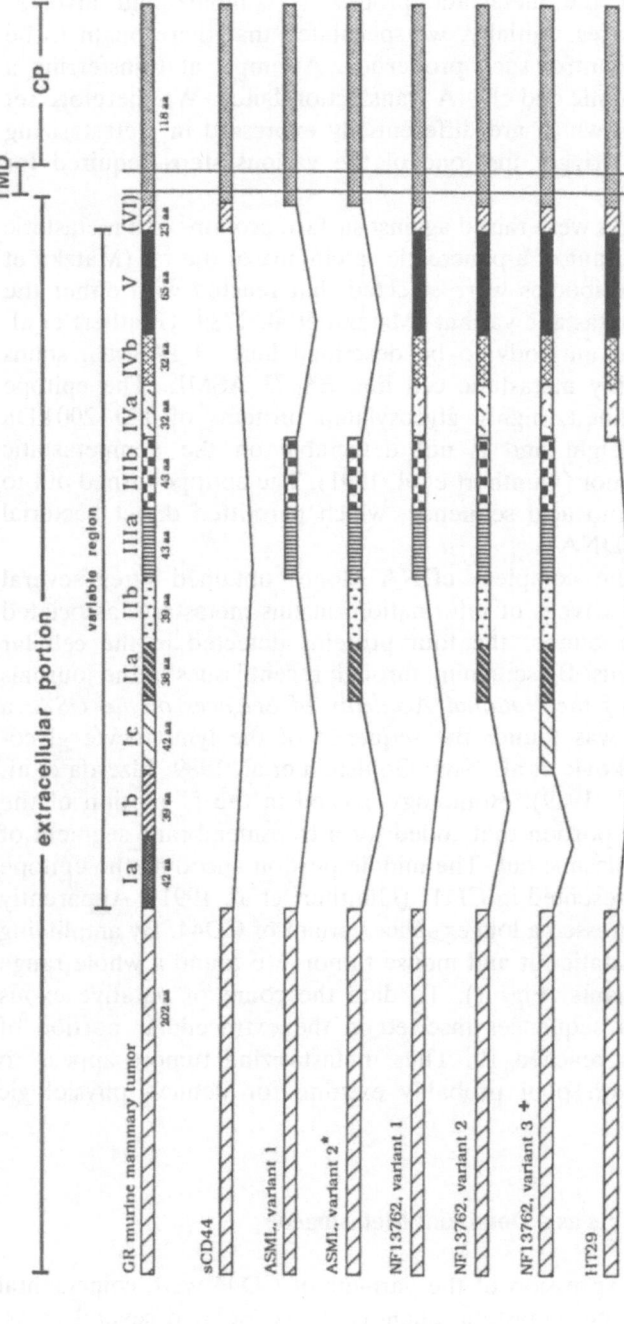

Fig. 1. Major splice variants of vCD44. *GR*, murine mammary tumor line derived from the GR mouse. For other cell lines see Günthert et al. (1991) and Hofmann et al. (1991). * Identical with meta 1 (Günthert et al. 1991); + identical with keratinocyte type (Hofmann et al. 1991)

CD44 cDNA fused to the SV40 or CMV promoter. These transfectants express one variant of CD44 in quantities comparable to or below that in the metastatic BSp73 ASML cells. The transfectants yielded a surprising result. While the recipient cells hardly ever reached the lymph nodes upon subcutaneous injection and never formed lung colonies (0/40), the transfectants were just as metastatic as the highly metastatic variant line (Günthert et al. 1991). All of the animals (60/60) receiving the SV40-driven clones developed lung metastases within 5–8 weeks. The CMV-driven clone produced metastases in five of six animals. Similar expression levels of the lymphocyte form of CD44 (sCD44) could not confer· metastatic behavior. Thus, the expression of the metastasis-specific variant appears to confer the complete metastatic behavior to a locally growing tumor cell.

Since the recipient tumor cell used for the transfection experiment was closely related to the highly metastatic tumor cell the gene was cloned from, the simplest interpretation of the results would be that from the complete genetic program required for metastasis formation the two cell lines differ just in one property, namely the expression of variant CD44. Alternatively the variant CD44 molecule could act as a pleiotropic effector on the metastasizing cells triggering the expression of part or all of the genetic metastasis program. If this interpretion were correct, other locally growing tumor cells without any relation to the pancreatic carcinoma cell lines would also become metastatic upon transfection of a vCD44 expression vector. This indeed appears to be the case. Fibrosarcoma tumor cells isolated from the same rat strain as the BSp73 tumors and rat 2 fibroblasts transformed with the activated Ha-*ras* gene both give rise to locally growing tumors upon subcutaneous injection. vCD44-expressing transfectants grow not only as primary tumors but metastasize to lymph nodes and lung. We therefore conclude that vCD44 contributes not just one function to the metastatic phenotype but acts as a pleiotropic regulator. In analogy to the oncogene concept, such a regulator would be called a metastogene.

The monoclonal antibody that was used for the identification and isolation of vCD44 was examined for its ability to interfere with metastatic spread. Animals were injected subcutaneously with tumor cells and at the same time intravenously with the monoclonal antibody. The injection of antibodies was repeated twice weekly. Primary tumors were removed by surgery at day 10. When the highly metastatic ASML cells were used, spread of metastases was retarded. The survival time of the animals was prolonged from 35 days to more than 70 days. When using the vCD44-transfected AS cells the effect of the antibody application was even more pronounced. About 70% of the animals treated by intravenous injection of the monoclonal antibody were permanently cured of tumor cells and survived. As there is no evidence for cytotoxic T cell recruitment or for an anti-idiotypic response, a block of the function of vCD44 by the binding of the monoclonal antibody is the most likely interpretation. Obviously, cells attempting to migrate to the lymph nodes cannot survive for longer periods of time unless they undergo a variant CD44-mediated interaction. Further, during migration tumor cells appear to be accessible to antibodies. This is an important finding in considering targeting in vivo.

Human Tumor Cell Lines and Surgically Resected Tumors Express Variants of CD44

During screening RNA from human tumor cell lines, several RNA probes were detected that hybridize to the variant protein of the rat CD44 sequence. The tumor cell lines derived from human large cell lung carcinoma, breast cancer and colon cancer expressed several larger RNA species with portions homologous to the lymphocyte CD44 and to the variant sequences, very similar to the RNA isolated from the rat tumor line. From one of these cell lines, the human cDNA was cloned (Hofmann et al. 1991). Because the monoclonal antibody 1.1ASML does not recognize an epitope on human molecules, we set out to prepare antibodies to the human variant sequence. We expressed the human cDNA as a fusion with the bacterial glutathione-S-transferase gene. Using the bacterially expressed protein we obtained a polyclonal rabbit antiserum. This serum stains cells in frozen sections or

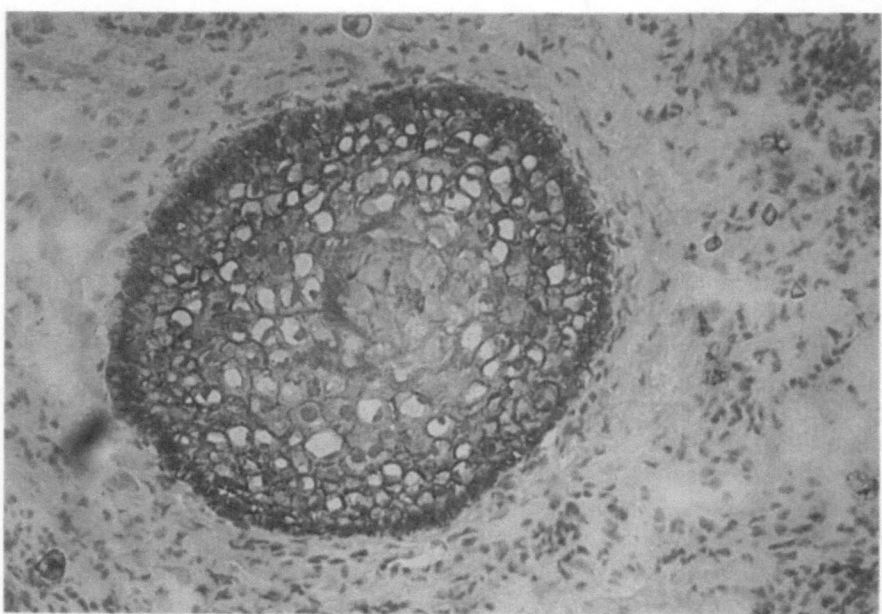

Fig. 2. Cryosection of a lung metastasis derived from a testis neoplasia. The section was fixed with methanol-acetone and incubated with affinity-purified anti-human vCD44 rabbit antiserum. Staining was performed using the avidin-biotin-peroxidase (ABC) technique

proteins in western blots. The preliminary screen through primary human tumor material has been very encouraging. Samples of resected lung metastases of various kinds of tumors showed heavily staining tumor cells. An example of a human testis neoplasia is shown (Fig. 2).

Summary and Perspectives

The locus encoding the variants of CD44 represents the first example of a pleiotropic gene (metastogene) that confers the full metastatic ability to a number of rat tumor cells. Antibodies recognizing the extracellular portion of the encoded transmembrane proteins effectively block the metastatic function. Our current interpretation of available data is that the protein serves as pleiotropic regulator either constitutively (e.g., by mutation) or in response to an unknown ligand. The function is required prior to metastatic colony formation in the draining lymph nodes. The initial results obtained with human cancer cells determine the direction of research in the near future. A survey of expression in a large series of human tumors will determine at what stage of tumor progression the metastogene is turned on. Further, the expression may define vCD44 as a prognosis factor. Equally important will be to search for the ligand to which the variant CD44 binds specifically and to explore the exact mechanism of its action.

References

Andres A-C, Schönenberger C-A, Groner B, Henninghausen L, LeMeur M, Gerlinger P (1987) Ha-ras oncogene expression directed by a milk protein gene promoter: tissue specificity, hormonal regulation, and tumor induction in transgenic mice. Proc Natl Acad Sci USA 84:1299–1303

Bailleul B, Surani MA, White S, Barton SC, Brown K, Blessing M, Jorcano J, Balmain A (1990) Skin hyperkeratosis and papilloma formation in transgenic mice expressing a ras oncogene from a suprabasal keratin promoter. Cell 62:697–708

Behrens J, Mareel MM, van Roy FM, Birchmeier W (1989) Dissecting tumor cell invasion: epithelial cells acquire invasive properties after the loss of uvomorulin cell-cell adhesion. J Cell Biol 198:2435–2447

Bernards R, Schrier PI, Houweling A, Bos JL, van der Eb AJ, Zijlstra M, Melief CJM (1983) Tumorigenicity of cells transformed by adenovirus type 12 by evasion of T-cell immunity. Nature 305:776–779

Chan BMC, Matsuura N, Takada Y, Zetter BR, Hemler ME (1991) In vitro and in vivo consequences of VLA-2 expression in rhabdomyosarcoma cells. Science 251:1600–1602

Dano K, Andreasen PA, Grondahl-Hansen J, Kristensen P, Nielsen LS, Skriver L (1985) Plasminogen activators, tissue degradation and cancer. Adv Cancer Res 44:139–266

Duijvestijn A, Hamann A (1989) Mechanisms and regulation of lymphocyte migration. Immunol Today 10:23–29

Giancotti FG, Ruoslathi E (1990) Elevated levels of the α5β1 fibronectin receptor suppress the transformed phenotype of Chinese hamster ovary cells. Cell 60:849–859

Goldstein LA, Zhou DFH, Picker LJ, Minty CN, Bargatze RF, Ding JF, Butcher EC (1989) A human lymphocyte homing receptor, the hermes antigen, is related to cartilage proteoglycan core and link proteins. Cell 56:1063–1072

Günthert U, Hofmann M, Rudy W, Reber S, Zöller M, Haußmann I, Matzku S, Wenzel A, Ponta H, Herrlich P (1991) A new variant of glycoprotein CD44 confers metastatic potential to rat carcinoma cells. Cell 65:13–24

Hanahan D (1985) Heritable formation of pancreatic β-cell tumors in transgenic mice expressing recombinant insulin/simian/virus 40 oncogenes. Nature 315:115–122

Hartveit E (1990) Attenuated cells in breast stroma: the missing lymphatic system of the breast. Histopathology 16:533–543

Hofmann M, Rudy W, Zöller M, Tölg C, Ponta H, Herrlich P, Günthert U (1991) CD44 splice variants confer metastatic behavior in rats: homologous sequences are expressed in human tumor cell lines. Cancer Res 51:5292–5297

Idzerda RL, Carter WG, Nottenburg C, Wayner EA, Gallatin WM, St, John T (1989) Isolation and DNA sequence of a cDNA clone encoding a lymphocyte adhesion receptor for high endothelium. Proc Natl Acad Sci USA 86:4659–4663

Kieran MW, Blank V, le Bail O, Israel A (1989) Lymphocyte homing. Res Immunol 140:399–450

Klein CE, Steinmayer T, Kaufmann D, Weber L, Brocker EB (1991) Identification of a melanoma progression antigen as integrin VLA-2. J Invest Dermatol 96:281–284

Matrisian LM, Bowden GT, Krieg P, Fürstenberger G, Briand J-P, Leroy P, Breathnach R (1986) The mRNA coding for the secreted protease transin is expressed more abundantly in malignant than in benign tumors. Proc Natl Acad Sci USA 83:9413–9417

Matzku S, Komitowski D, Mildenberger M, Zöller M (1983) Characterization of BSp73, a spontaneous rat tumor, and its in vivo selected variants showing different metastasizing capacities. Invest Metastasis 3:109–123

Matzku S, Wenzel A, Liu S, Zöller M (1989) Antigenic differences between metastatic and nonmetastatic BSp73 rat tumor variants characterized by monoclonal antibodies. Cancer Res 49:1294–1299

Nicolson GL (1988) Cancer metastasis: tumor cell and host organ properties important in metastasis to specific secondary sites. Biochim Biophys Acta 948:175–224

Nottenburg C, Rees G, St. John T (1989) Isolation of mouse CD44 cDNA: structural features are distinct from the primate cDNA. Proc Natl Acad Sci USA 86:8521–8525

Ossowski L, Reich E (1983) Antibodies to plasminogen activator inhibit human tumor metastasis. Cell 35:611–619

Pauli BU, Augustin-Voss HG, El-Sabban ME, Johnson RC, Hammer DA (1990) Organ-preference of metastasis. The role of endothelial cell adhesion molecules. Cancer Metastasis Rev 9:175–190

Plaksin D, Gelber C, Feldman M, Eisenbach L (1988) Reversal of the metastatic phenotype in Lewis lung carcinoma cells after transfection with syngeneic H-2K[b] gene. Proc Natl Acad Sci USA 85:4463–4467

Plantefaber LC, Hynes RO (1989) Changes in integrin receptors on oncogenically transformed cells. Cell 56:281–290

Raz A, Lotan R (1987) Endogenous galactoside-binding lectins: a new class of functional tumor cell surface molecules related to metastasis. Cancer Metastasis Rev 6:433–452

Schoenenberger C-A, Andres A-C, Groner B, van der Valk M, LeMeur M, Gerlinger P (1988) Targeted c-myc gene expression in mammary glands of transgenic mice induces mammary tumors with constitutive milk protein gene transcription. EMBO J 7:169–175

Schrier PI, Bernards R, Vaessen RTMJ, Houweling A, van der Eb AJ (1983) Expression of class I major histocompatibility antigens switched off by highly oncogenic adenovirus 12 in transformed rat cells. Nature 305:771–775

Sinn E, Muller W, Pattengale P, Tepler I, Wallace R, Leder P (1987) Coexpression of MMTV/v-Ha-ras and MMTV/c-myc genes in transgenic mice: synergistic action of oncogenes in vivo. Cell 49:465–475

Stamenkovic I, Amiot M, Pesando JM, Seed B (1989) A lymphocyte molecule implicated in lymph node homing is a member of the cartilage link protein family. Cell 56:1057–1062

Stewart TA, Pattengale PK, Leder P (1984) Spontaneous mammary adenocarcinomas in transgenic mice that carry and express MTV/myc fusion genes. Cell 38:627–637

Stoolman LM (1989) Adhesion molecules controlling lymphocyte migration. Cell 56:907–910.

Tryggvason K, Höyhtyä M, Salo T (1987) Proteolytic degradation of extracellular matrix in tumor invasion. Biochim Biophys Acta 907:191–217

Wallich R, Bulbuc N, Hämmerling GJ, Katzav S, Segal S, Feldman M (1985) Abrogation of metastatic properties of tumour cells by de novo expression of H-2K antigens following H-2 gene transfection. Nature 315:301–305

Progression in Human Melanoma Is Accompanied by the Altered Expression of Cell Adhesion Molecules*

J.P. Johnson, B.G. Stade, U. Rothbächer, S. Stratil, and G. Riethmüller

Institut für Immunologie, Goethestraße 31, W-8000 München 2, FRG

Introduction

Alterations in cell-cell interactions are characteristic of malignant cells. Normal interactions with neighboring cells are disturbed as the tumor cells invade the surrounding connective tissue, traverse the vascular and lymphatic systems and take up residence in foreign environments. Many of the steps in this process could reflect changes in the cell adhesion molecules expressed by the tumor cells. Loss of normal cell adhesion is likely to be one of the earliest events in the metastatic cascade, not only allowing individual tumor cells to separate from the tumor mass but also reducing numerous contact-mediated controls [9, 23]. However, tumor cells not only demonstrate a loss of normal cellular interactions, they also display new interactions with elements of the vasculature and with cells at the sites of secondary growths. These new interactions could, at least in part, also be due to changes in cell adhesion molecules – in this case newly expressed molecules mediating heterotypic adhesion. In this context it is striking that two melanoma-associated antigens, identified solely from the observation that they first become detectable as the tumors develop metastatic capacity, have turned out to be cell adhesion molecules which likely mediate heterotypic cell adhesion.

Progression in Cutaneous Melanoma

The development of human malignant melanoma, like that of many other human and experimental animal tumors, is a complex multistep process [31]. Because of the pigmented nature of the melanocytes and their location in

* This work was supported by grants from the Deutsche Krebshilfe, Mildred Scheel Stiftung and the Deutsche Forschungsgemeinschaft, SFB 217(A3) Bonn.

Wagener/Neumann (Eds.)
Molecular Diagnostics of Cancer
© Springer-Verlag Berlin Heidelberg 1993

the skin, this process can be observed in the broad spectrum of melanocytic lesions which have been identified by histopathologists [5]. While normal melanocytes in the epidermis are found dispersed singly along the dermal-epidermal border, a proliferation of these cells gives rise to melanocytic nevi or moles. Although these melanocytic tumors are benign, the presence of large numbers of them is nevertheless associated with an increased risk for melanoma [18]. Dysplastic nevi demonstrate both cytologic and architectural atypia, and because of their high incidence in members of melanoma-prone families they are considered by many histopathologists to be premalignant lesions [10]. Among primary melanomas, the vertical thickness of the tumor is one of the best prognostic parameters [4]. Thus the 5-year survival rate for patients with tumors < 0.75 mm in thickness is approximately 98%, while for patients with tumors > 3.00 mm in thickness it is less than 60% [18]. This correlation is a reflection of a change in the tumor from a radial growth phase into a vertical growth phase, a change which has been shown to be associated with the development of metastatic potential [6]. While the histological and clinical characteristics of these distinct melanocytic lesions have been well studied, almost nothing is known about the changes in gene expression which characterize them. Identifying such molecular changes would not only help us to understand how melanoma develops and progresses to metastatic disease, but could lead to the identification of markers for metastatic cells which could have important prognostic and therapeutic applications.

Identification of Progression Markers in Melanoma

To identify molecules whose expression may correlate with the development of metastatic potential in melanoma, we have tried to isolate monoclonal antibodies (MAbs) which show a differential reactivity with melanocytes present in the benign, malignant and metastatic lesions. Mice were immunized with melanoma metastases and the growing hybridomas were examined for their reactivity with frozen tissue sections of metastases and melanocytic nevi. Figure 1 presents a summary of the reactivity of 10 MAbs which we have obtained using this procedure. The gray columns indicate the frequency of reactive melanomas (> 100 primaries and metastases tested) while the black columns indicate the frequency of reactive nevi (> 50). The majority of melanoma-associated antigens which have been defined to date are expressed in a qualitatively similar way by both benign and malignant tumors, as exemplified by the melanoma chondroitin sulfate proteoglycan (CSP [32]). However, using the protocol detailed above, antibodies have been obtained which stain most metastatic lesions but which are only rarely reactive with benign melanocytic tumors. Biochemical analyses of the antigens isolated from melanoma cell lines indicate that most, although not all [24], of the molecules defined to date are proteins. Examination of

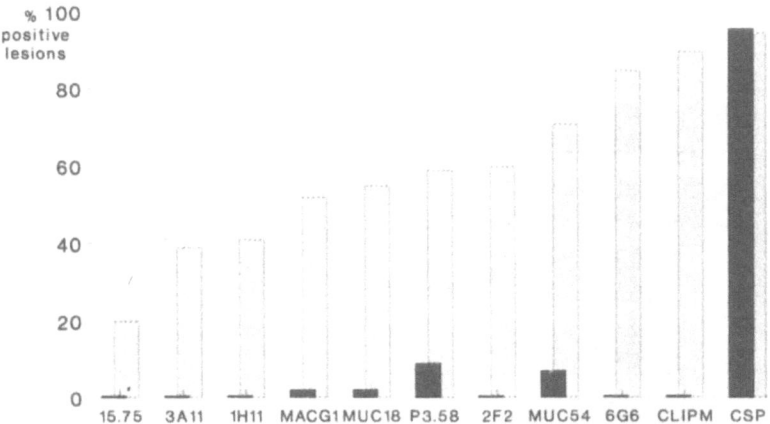

Fig. 1. Comparison of antibody reactivity on malignant melanomas (>100) and melanocytic nevi (>50). Data are shown as frequency of positive lesions and are taken from immunoperoxidase staining of frozen sections of melanomas (*gray bars*) and nevi (*black bars*). The names on the *x* axis denote the antibody. *CSP*, melanoma chondroitin sulfate antigen

the reactivity of the antibodies with nonmelanocytic normal and malignant tissues indicates that each antibody has a unique tissue distribution. Some such as CLIP.M are broadly expressed on normal and malignant tissues while others such as MUC18 are only rarely found on nonmelanoma tissues.

When the antibodies are tested for their reactivity with primary malignant melanomas staged according to their vertical thickness, they can be divided into two groups (Fig. 2). The first group, exemplified by MUC54, CLIP.M and MACG1, react with most malignant lesions regardless of the stage. Such antibodies identify molecules which we have called early progression antigens. These molecules are only rarely detectable on benign melanocytic lesions but are strongly expressed on malignant lesions of all stages and may be involved in the transformation of the melanocyte. The second group of antibodies, exemplified by P3.58, MUC18 and 15.75 (as well as 6G6, 1H11 and 3A11), show a different type of pattern. These antibodies are strongly reactive with metastatic lesions, and among the primary tumors they demonstrate a correlation with the vertical thickness. The molecules defined by these antibodies are known as late progression markers, since they are only rarely expressed by the very thin primary tumors which have a low probability of metastasis but are characteristic of the more advanced primary tumors and metastases. Such antigens are candidates for molecules which might contribute to the development of metastases.

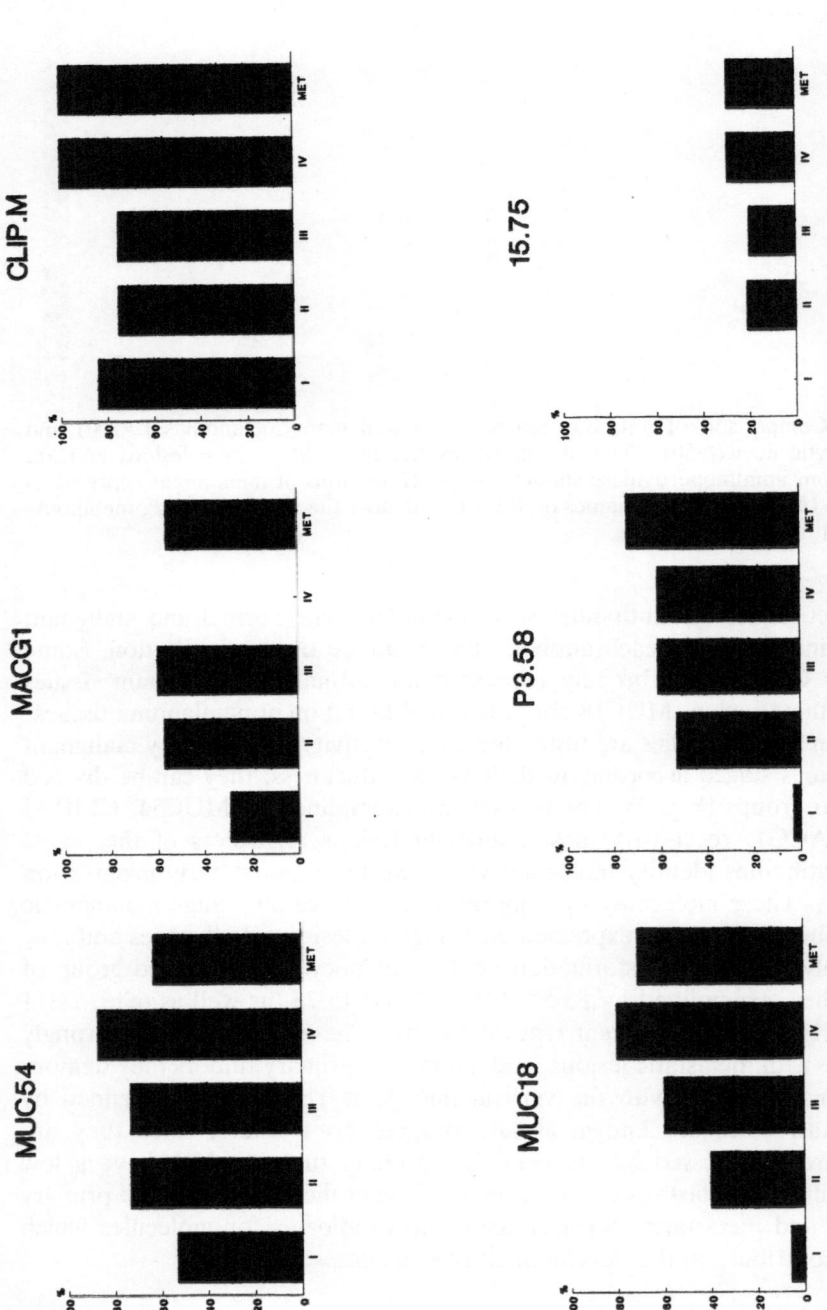

Fig. 2. Antibody reactivity with malignant melanomas. Primary melanomas are divided according to vertical thickness: *I*, < 0.75 mm; *II*, 0.76–1.50 mm; *III*, 1.51–3.00 mm; *IV*, > 3.00 mm; *MET*, metastasis. MACG1 was not tested on group IV primary tumors

Characterization of Two Late Progression Markers in Melanoma

The late progression antigens defined by MAbs P3.58 and MUC18 show a similar pattern of expression on melanocytic lesions in situ [12]. Both are only rarely observed on nevi but are characteristic for metastases. The thin primary tumors, which have a low probability to metastasize, are no more frequently positive for these markers than are the benign lesions. Expression of P3.58 and MUC18 first becomes significant on tumors which have reached a vertical thickness of approximately 1 mm. With increasing tumor thickness the frequency of MUC18 expression increases while that of P3.58 remains relatively stable. In both cases the fraction of positive cells per tumor shows an increase with increasing tumor thickness. Such a pattern of expression (sporadic in tumors with no metastatic potential and increasingly strong in those with increasing metastatic potential) is consistent with the hypothesis that these molecules contribute to the development of metastatic potential, and it was therefore important to try to obtain an idea of what their functions might be. One approach to this problem is to obtain the cDNAs encoding these molecules. From a comparison with the sequences of molecules with proven functional domains, it is frequently possible to predict a new molecule's function. In any case, the availability of the cDNA makes it possible via transfection to specifically express the antigen in cells lacking it and to look for functional changes in the cell. The cDNA encoding both of these markers was obtained by screening melanoma cDNA lambda expression libraries with second-round MAbs produced against the denatured antigen isolated by affinity chromatography from melanoma cells.

The P3.58 Antigen Is Identical to Intercellular Adhesion Molecule 1 (ICAM-1)

The P3.58 antigen is a cell surface glycoprotein with an apparent molecular weight of 89 kDa in melanoma cells [13]. It was found to be expressed on some blood vessel endothelia, in germinal centers of lymph follicles and on mitogen-activated peripheral blood lymphocytes, indicating that its normal function may be in the immune system [14]. Antibodies directed to the antigen were in fact able to block the formation of lymphoid cell clusters in vitro [15], suggesting that the antigen participates in leukocyte cell-cell adhesion.

Three overlapping clones were isolated from the cDNA libraries and their sequencing was about 70% completed when the sequence of ICAM-1 was published [26]. The sequence of P3.58 antigen isolated from melanoma cells [17] is identical to that of ICAM-1 isolated from myeloid cells [26] or endothelial cells [28].

MUC18 Is a Novel Member of the Ig Superfamily Cell Adhesion Molecules

The MUC18 antigen is a single-chain cell surface glycoprotein with an apparent molecular weight of 113 kDa [19]. Examination of frozen sections of a wide range of normal and malignant tissues indicated that the expression of MUC18 in normal adults is highly restricted to smooth muscle in some blood vessels [19]. Sequencing of MUC18-encoding cDNA clones revealed that MUC18 has all the characteristics of an integral membrane protein and that it is a member of the immunoglobulin supergene family [20]. Although MUC18 is a novel member of this family, it is related to several other molecules, all of which have been shown to mediate intercellular adhesion (Table 1). The highest sequence similarity is to DCC, a putative tumor suppressor gene in colorectal carcinoma [9]. This molecule, whose gene is mutated or deleted in most of these tumors, has recently been shown to mediate homotypic cell adhesion. MUC18 is also related to the carcinoembryonic antigen family, several members of which have been shown in transfection studies to mediate heterotypic and/or homotypic adhesion [2, 22] and to a group of molecules (NCAM, L1, MAG) mediating cell-cell adhesion during the development of the nervous system [8]. Because of the sequence similarities to these molecules, it seems likely that MUC18 also is a cell adhesion molecule.

A Role for MUC18 and ICAM-1 in Melanoma Progression?

While the function of MUC18 remains unknown, its expression on smooth muscle cells of some blood vessels suggests that its ligand might be found in the vasculature. If this is true, then it is tempting to speculate that the expression of MUC18 by melanomas may in some way promote intra- and extravasation of the tumor cells.

Table 1. Proteins showing sequence similarity to the MUC18 glycoprotein

Molecule	Percentage identity	Amino acid overlap	Optimized score	Value
DCC	22.5	377	195	25
CEA	20.2	297	167	20.8
PSGP	18.4	288	168	20.9
NCA	20.5	273	160	19.7
NCAM	21.9	196	167	20.8
L1	22.6	288	172	22
Amalgam	26.0	169	133	15.5
MAG	20.8	448	135	15.8
Fasciclin II	16.2	229	132	15.4
Contactin	22.2	117	79	13.3

Amino acid sequence comparisons were performed with the FASTP program. A z value of > 10 is considered significant.

The known ligands of ICAM-1 are LFA-1 (CD11a), expressed by most leukocytes, and MAC-1 (CD11b), characteristic of myeloid cells and granulocytes. ICAM-1 therefore mediates the adhesion of cells expressing it to leukocytes; its primary function appears to be to strengthen interactions between immune cells and their targets [27]. The expression of ICAM-1 by a tumor cell would therefore be expected to enhance its recognition and killing by specific and nonspecific effector cells and to be, if anything, associated with a good prognosis. Consistent with this is the observation that in B-cell lymphomas [11] and in renal cell carcinomas [29] ICAM-1 expression is associated with the highly differentiated tumors which have the best prognosis. In contrast, however, the expression of ICAM-1 by melanomas is associated with advanced tumors which have a poor prognosis. Not only was ICAM-1 identified as a late progression marker in this tumor [16], but a prospective study of stage I melanomas using an independently generated antibody has shown that patients whose tumors expressed ICAM-1 have a significantly shorter disease-free interval than patients with ICAM-1-negative tumors [21].

In vitro studies have shown that ICAM-1 functions to enhance CTL and LAK killing of melanoma cells as it does in other types of tumors [1]. Nevertheless it remains possible that the function of ICAM-1 on melanomas is in some way altered. ICAM-1 isolated from melanoma cells is more highly glycosylated than that from autologous B cells [14] and may therefore differ in particular aspects of ligand binding [7]. In addition, a soluble form of the molecule [25] can be detected in culture supernatants from melanoma cell lines, raising the possibility that it is also produced in vivo and inhibits local immune cell–tumor interactions. It is not, however, necessary to invoke functional alterations in ICAM-1 to postulate a role for this molecule in tumor progression. Numerous studies in animal models have shown that aggregates of tumor cells and leukocytes are more effective than tumor cells alone in generating spontaneous as well as experimental metastases [3]. Rather than resulting in destruction of the tumor cells, the presence of leukocytes more often enhances invasion of the extracellular matrix, provides local growth and angiogenesis factors, and promotes the intra-/extravasation of the tumor cells. The expression of ICAM-1 by melanoma cells has been shown to enhance their binding to leukocytes in vitro. In vivo, ICAM-1 expression may also increase the tendency for tumor–leukocyte aggregates to form and thus enhance the probability of metastasis formation. In any case, the evidence that either MUC18 or ICAM-1 actually contributes to metastasis development must await more direct tests in animal models [30].

References

1. Anchini A, Mortarini R, Supino R, Parmiani G (1990) Human melanoma cells with high susceptibility to cell-mediated lysis can be identified on the basis of ICAM-1 phenotype, VLA profile and invasive ability. Int J Cancer 46:508–515

2. Benchimol, S, Fuks A, Jothy S, Beauchemin N, Shirota K, Stanners CD (1989) Carcinoembryonic antigen, a human tumor marker, functions as an intercellular adhesion molecule. Cell 57:327–334

3. Blood CH, Zetter BR (1990) Tumor interactions with the vasculature: angiogenesis and tumor metastasis. Biochim. Biophys. Acta 1032:89–118

4. Breslow, A (1970) Thickness, cross-sectional areas and depth of invasion in the prognosis of cutaneous melanoma. Ann Surg 172:902–908

5. Clark WH, Elder DE, Guerry D, Epstein MN, Greene MH, van Horn M (1984) A study of tumor progression. The precursor lesions of superficial spreading and nodular melanoma. Hum Pathol 15:1147–1165

6. Clark WH, Elder DE, Guerry D, Braitman LE, Trock BJ, Schultz D, Synnestvedt M, Halpern AC (1989) Model predicting survival in stage I melanoma based on tumor progression. J Natl Cancer Inst 81:1893–1904

7. Diamond MS, Staunton DE, Marlin SD, Springer TA (1991) Binding of the integrin Mac-1 (CD11b/CD18) to the third immunoglobulin-like domain of ICAM-1 (CD54) and its regulation by glycosylation. Cell 65:961–971

8. Edelman GM, Crossin KL (1991) Cell adhesion molecules: implications for a molecular histology. Annu Rev Biochem 60:155–190

9. Fearon ER, Cho KR, Nigro JM, Kern SE, Simons JW, Ruppert JM, Hamilton SR, Preisinger AC, Thomas G, Kinzler KW, Vogelstein B (1990) Identification of a chromosome 18q gene that is altered in colorectal cancers. Science 247:49–56

10. Greene MH, Clark WH, Tucker MA, Elder DE, Kraemer KH, Guerry D, Witmer WK, Thompson J, Matozzo I, Fraser MC (1985) Acquired precursors of cutaneous malignant melanoma. The familial dysplastic nevus syndrome. N Engl J Med 312:91–94

11. Gregory CD, Murray RJ, Edwards CF, Rickinson AB (1988) Downregulation of cell adhesion molecules LFA-3 and ICAM-1 in Epstein-Barr virus positive Burkitt's lymphoma underlies tumor cell escape from virus specific T cell surveillance. J Exp Med 167:1811–1824.

12. Holzmann B , Bröcker EB, Lehmann JM, Ruiter DJ, Sorg C, Riethmüller G, Johnson JP (1987) Tumor progression in human melanoma: five stages defined by their antigenic phenotypes. Int J Cancer 39:466–471

13. Holzmann B, Johnson JP, Kaudewitz P, Riethmüller G (1985) In situ analysis of antigens on malignant and benign cells of the melanocyte lineage: differential expression of 2 surface molecules, gp75 and gp89. J Exp Med 161:366–377

14. Holzmann B, Lehmann JM, Ziegler-Heitbrock HW, Funke I, Riethmüller G, Johnson JP (1988) Glycoprotein p3.58, associated with tumor progression in malignant melanoma, is a novel leukocyte activation antigen. Int J Cancer 41:542–547

15. Johnson JP, Lehmann JM, Riethmüller G (1988) The progression associated human melanoma antigen p3.58 mediates monocyte-lymphocyte interactions in vitro. Eur J Immunol 18:2097–2100

16. Johnson JP, Stade BG, Holzmann B, Schwäble W, Riethmüller G (1989) De novo expression of cell adhesion molecule ICAM-1 in melanoma and increased risk of metastasis. Proc. Natl Acad Sci USA 86:641–644

17. Johnson JP, Stade BG, Hupke U, Holzmann B, Riethmüller G (1988) The melanoma progression associated antigen p3.58 is identical to the intercellular adhesion molecule ICAM-1. Immunobiology 178:275–284

18. Koh HK (1991) Cutaneous melanoma. N Engl J Med 325:171–182

19. Lehmann JM, Holzmann B, Breitbart EW, Schmiegelow P, Riethmüller G, Johnson JP (1987) Discrimination between benign and malignant cells of the melanocytis lineage by two novel antigens, a glycoprotein with a molecular weight of 113 000 and a protein with a molecular weight of 76 000. Cancer Res. 47:841–847

20. Lehmann JM, Riethmüller G, Johnson JP (1989) MUC18, a marker of tumor progression in human melanoma, shows sequence similarity to the neural cell adhesion molecules of the immunoglobulin superfamily. Proc Natl Acad Sci USA 86:9891–9895

21. Natali P, Nicotra MR, Cavaliere R, Bigotti A, Romano G, Temponi M, Ferrone S (1990) Differential expression of intercellular adhesion molecule 1 in primary and metastatic melanoma lesions. Cancer Res 50:1271–1278

22. Oikaea S, Inuzuka C, Kuroki M, Matsuok Y, Kosati G, Nakazato H (1989) Cell adhesion activity of nonspecific cross reactive antigen (NCA) and carcino-embryonic antigen (CEA) expressed on CHO cell surfaces: homophilic and heterophilic adhesion. Biochem Biophys Res Commun 164:39–45

23. Schipper J, Frixen U, Behrens J, Unger A, Jahnke K, Birchmier W (1991) E-cadherin expression in squamous cell carcinomas of head and neck: inverse correlation with tumor dedifferentiation and lymph node metastases. Cancer Res 51:6328–6337

24. Schriever F, Dennis RD, Riethmüller G, Johnson JP (1988) MACG1, a mouse monoclonal antibody detecting a monosialoganglioside expressed in tumor-infiltrating macrophages. Cancer Res 48:2524–2530

25. Seth R, Raymond FD, Makgoba MW (1991) Circulating ICAM-1 isoforms: diagnostic prospects for inflammatory and immune disorders. Lancet 338:83–84

26. Simmons D, Makgoba MW, Seed B (1988) ICAM-1, an adhesion ligand of LFA-1, is homologous to the neural cell adhesion molecule NCAM. Nature 331:624–627

27. Springer TA (1990) Adhesion receptors of the immune system. Nature 346:425–434

28. Staunton DE, Marlin SD, Stratowa C, Dustin ML, Springer TA (1988) Primary structure of ICAM-1 demonstrates interaction between members of the immunoglobulin and integrin supergene families. Cell 52:925–933

29. Tomita Y, Nishiyama T, Watanabe H, Fujiwara M, Sato S (1990) Expression of intercellular adhesion molecule-1 (ICAM-1) on renal-cell cancer: possible significance in host immune responses. Int J Cancer 46:1001–1006

30. van Muijen, GNP, Cornelissen LMHA, Jansen CFJ, Figdor CG, Johnson JP, Bröcker EB,   Ruiter DJ (1991) Antigen expression of metastasizing and non-metastasizing human melanoma cells xenografted into nude mice. Clin Exp Metastasis 9:259–272

31. Weinberg RA (1989) Oncogenes, antioncogenes and the molecular basis of multistep carcinogenesis. Cancer Res 49:3713–3721

32. Wilson BS, Imai K, Natali PG, Ferrone S (1981) distribution and molecular characterization of a cell surface and a cytoplasmic antigen detectable in human melanoma cells with monoclonal antibodies. Int J Cancer 28:293–300

Structure, Function and Expression of the CEA Gene Family: Diagnostic and Therapeutic Implications

W. Zimmermann, F. Grunert, G. Nagel, S. von Kleist, and J. Thompson

Institut für Immunbiologie, Universität Freiburg, Stefan-Meier-Straße 8, W-7800 Freiburg, FRG

Introduction

During the search for biochemical differences between tumorous and corresponding normal tissues a quarter of a century ago, the carcinoembryonic antigen (CEA) was found in colonic tumors, but not in normal colonic mucosa (Gold and Freedman 1965; von Kleist and Burtin 1966). With the development of more sensitive analytical methods, CEA and most other "tumor-specific" markers were also detected in normal tissues or in sera of individuals without tumors (for review see Shively and Beatty 1985). Despite this fact, CEA is widely used for the monitoring of tumor patients for recurrence of malignant disease after surgery (Fantini and DeCosse 1990). A continuous rise in the CEA concentration in serum, detected by serial determinations, is an indicator of tumor regrowth or metastasis in patients with adenocarcinomas of the colon, rectum, breast, lung and pancreas. Since an increase of the CEA concentration is often observed before other clinical symptoms are obvious, early therapeutic measures can be taken (e.g., "second-look" surgery in patients with colorectal tumors (Fantini and DeCosse 1990)). However, the diagnostic value of CEA, e.g., for early detection of primary tumors by routine screening, is limited due to the low sensitivity and specificity of CEA measurements. Therefore, in general, only patients with advanced malignant disease show increased preoperative CEA serum concentrations. On the other hand, elevated CEA concentrations can also be detected in patients with benign disease (e.g., colitis) and in smokers (Shively and Beatty 1985). Clinical trials indicate that radioactively labeled antibodies directed against CEA can be used to localize primary tumors and metastases (Bischoff-Delaloye et al. 1989). Due to the availability of other extremely potent imaging methods, such as computed tomography and magnetic resonance imaging, immunoscintigraphy will probably not be used routinely in the future. However, another immunolocalization technique might soon become invaluable.

Wagener/Neumann (Eds.)
Molecular Diagnostics of Cancer
© Springer-Verlag Berlin Heidelberg 1993

Currently, the use of fluorescent dye-labeled, anti-CEA antibodies for the intraoperative detection of small tumors (e.g., tumor cells in regional lymph nodes) is being evaluated in athymic mice carrying human colonic carcinoma xenografts (Pelgrin et al. 1991). Furthermore, it is hoped that by the application of toxin- or radioactive isotope-coupled anti-CEA antibodies, the specificity of tumor therapy will be increased. Tumor regression has been observed in animal models by several groups using this approach (Buchegger et al. 1988; Sharkey et al. 1991). In the athymic mouse model, enhanced tumor localization and radioimmunotherapy by anti-CEA antibodies was achieved after administration of cytokines, which selectively stimulate CEA expression (Kuhn et al. 1991).

Besides CEA, a large family of closely related cross-reacting antigens have been described which differ in size and tissue distribution (reviewed in Thompson and Zimmermann 1988). For example, nonspecific cross-reacting antigen (NCA) is found in many tumors of epithelial origin, normal lung and spleen and polymorphonuclear cells (Bordes et al. 1975). Biliary glycoprotein, on the other hand, is expressed in hepatocellular carcinomas and normal epithelial cells of bile canaliculi (Svenberg 1976; Hinoda et al. 1990). The presence of CEA-cross-reacting antigens in normal tissues can interfere with measurement in sera and targeting on tumor cells of CEA by antibodies. Due to the high degree of glycosylation of this protein family, biochemical characterization proved to be very difficult. It was hoped, therefore, that cloning of the CEA gene and possibly of related genes would help to clarify the relationship of this complex protein family. This approach was also expected to yield information that would allow the production of more specific probes for diagnosis and therapy. Furthermore, using these probes for determination of the expression pattern of the various CEA family members, new tumor markers might be identified.

Structure, Function and Expression of the CEA Gene Family

Genomic and cDNA cloning have revealed that CEA and related antigens are encoded by a family of genes which belong to the immunoglobulin superfamily. To date 22 genes have been identified (Table 1). Based on sequence similarity they can be subdivided into two main subgroups: the CEA subgroup, which contains the CEA gene and the genes for the classical cross-reacting antigens, and the pregnancy-specific glycoprotein (PSG) gene subgroup. The latter group of genes code for highly similar proteins formerly not known to be related to CEA. PSGs are produced in large amounts in the fetal part of the placenta and secreted into the maternal blood. At term, PSGs comprise the most abundant placental proteins in sera of pregnant women.

The deduced primary structure reveals that the CEA-related antigens are composed of a leader peptide, which is removed after transport into the

Table 1. The presently known members of the CEA gene family

CEA subgroup	PSG subgroup
CEAa,b	PSG1a,b,c,d,e,f
NCA	PSG2n
BGPa,b,c,d,e,f,g,h,i	PSG3m
CGM1a,b,c	PSG4a
CGM2	PSG5n,m
CGM6	PSG6r,s
CGM7	PSG7
CGM8[a]	PSG8a
CGM9	PSG11s,w
CGM10[a,b]	PSG12
CGM11[a,b]	PSG13
	PSG14
	PSG15

The lower case letters indicate the various splice or polyadenylation mRNA variants. CGM, CEA gene family member; PSG, pregnancy-specific glycoprotein.
[a] Probably a pseudogene.
[b] W. Khan and S. Hammarström, personal communication.

endoplasmic reticulum, one immunoglobulin variable (IgV)-like domain and a varying number (none, two, three or six) of Ig constant (C) region-like domains (review: Thompson et al. 1991). Most members of the CEA subgroup seem to be membrane-bound either via a glycosyl-phosphatidylinositol anchor or a transmembrane domain. PSGs lack hydrophobic domains, which is in agreement with their accumulation in the maternal blood. Differential splicing increases the complexity of the CEA family. Up to seven proteins can be predicted for a single gene (BGP) which differ in the number of Ig C region-like domains or the size of their cytoplasmic tails.

All members of the CEA gene family are located on the long arm of chromosome 19 (19q13.2–3; review: Thompson et al. 1991). Mapping aided by pulsed field gel electrophoresis and "contig" analyses (Branscomb et al. 1990) demonstrated arrangement of the CEA-related genes in clusters on a 1.2-Mb chromosome segment. The members of the CEA subgroup are located in two smaller clusters followed by the tightly clustered PSG genes toward the telomere (Thompson et al. 1992). The close vicinity of the members of the CEA family could allow coordinate expression of pairs or groups of genes by the use of common regulatory elements.

All members of the CEA subgroup so far analyzed are able to convey in vitro cell adhesion properties to transfectants expressing individual CEA-related cDNAs (Table 2). This property has also been reported for a number of other members of the immunoglobulin superfamily, such as neural cell adhesion molecule (N-CAM) and myelin-associated glycoprotein (MAG; Williams 1987). CEA, NCA and BGP allow both homophilic and heterophilic cell adhesion, whereas the CGM6 gene product interacts only heterophilically with NCA and not homophilically with itself (Table 2). Detailed histological studies on the localization of rat CEA-related antigens imply that some members are involved in intercellular adhesion while others might aid in the organization of microvilli (Öbrink 1991). Disturbances in the expression of CEA or related antigens in tumor cells might contribute to the malignant phenotype (Benchimol et al. 1989), as is assumed to be the case for the recently discovered recessive oncogene product DCC ("deleted in colonic cancer"), a presumed cell adhesion molecule (Fearon et al. 1990). As an additional function of CEA subgroup members, binding of entero-bacteria such as *Escherichia coli* and *Salmonella typhi* via lectin molecules on type 1 fimbriae has been reported (Table 2). This interaction, which involves D-mannosyl residues, might be important for the colonization by bacteria of the colonic mucosa as well as for the recognition of bacteria by granulocytes, which express, with the exception of CEA, all so far charac-terized members of the CEA subgroup (see below). Recently, the presumed mouse BGP homologue, mmCGM2, was reported to represent the receptor of the mouse coronavirus, which causes hepatitis (Williams et al. 1991). It might, therefore, be possible that human coronaviruses, which also cause common respiratory illnesses, use members of the human CEA family

Table 2. Function(s) of CEA family members

Member	Function(s)	Reference
CEA	Homo- and heterophilic cell adhesion (Ca^{2+}- and temperature-independent)	Benchimol et al. 1989 Oikawa et al. 1989
	Binding of bacteria	Leusch et al. 1990
	Accessory molecule for collagen type I binding	Pignatelli et al. 1990
NCA-50/90	Homo- and heterophilic cell adhesion (Ca^{2+}- and temperature-independent)	Oikawa et al. 1989 Zhou et al. 1990
	Binding of bacteria	Leusch et al. 1990
BGP	Homophilic cell adhesion (Ca^{2+}- and temperature-dependent)	Rojas et al. 1990
	Binding of bacteria	Leusch et al. 1991
NCA-95 (CGM6)	Heterophilic cell adhesion with NCA-50/90	Oikawa et al. 1991
Ecto-ATPase/ Cell-CAM 105 (rat)	Homophilic cell adhesion (Ca^{2+}-independent)	Lin and Giodotti 1989 Aurivillius et al. 1990
mmCGM1/2 (mouse)	Homophilic cell adhesion (Ca^{2+}- and temperature-dependent)	Turbide et al. 1991
	Mouse hepatitis virus receptor	Williams et al. 1991

as cell entry vehicles. The function of PSGs is unknown. Based on the inhibitory influence of PSGs on certain in vitro immunological reactions, it is speculated that they might be involved in the protection of the allotypic fetus from the maternal immune system by specific immunosuppression.

As a basis for improvement of the specificity and sensitivity of CEA detection, we have started to characterize the recognition pattern of a large panel of monoclonal antibodies. Among them are a number of anti-CEA monoclonal antibodies, the epitopes of which have been compared recently (Hammarström et al. 1989). To this end, we have tested a set of transfectants which express individual members of the CEA gene family (CEA, NCA, BGP, CGM1, CGM6) with each monoclonal antibody. After tagging with fluorescein-labeled second antibody, the antibody binding to the transfectants was determined by FACScan analyses. The results obtained by this approach also allow the assignment of biochemically characterized members of the CEA family to their respective genes by comparing reactivity patterns (Berling et al. 1990). After analysis of more than 110 monoclonal antibodies, three have been identified that react with only one transfectant each (CEA, NCA, CGM6). Two antibodies were found which react with the CEA transfectant and with the CGM1 or the BGP transfectant respectively. Since CEA is not found on granulocytes these antigens can be detected specifically on these cells and studied individually. The above-mentioned approach has also been used to characterize antibodies which define the clusters of differentiation (CD) 66 and 67 and have been shown to cross-react with CEA-related antigens. Whereas the CD67 antibody seems to be specific for the CGM6 product, the CD66 antibodies exhibit a broader recognition pattern, reacting with CEA, NCA, BGP and CGM1 (Watt et al. 1991 and unpublished results). Since CD66 and CD67 antibodies have been shown to react within the hematopoietic system exclusively with mature granulocytes and some precursors, CEA-related antigens therefore represent surface markers for the myeloid lineage.

As long as not all members of the CEA family can be discriminated by monoclonal antibodies, in parallel we have applied gene-specific hybridzation probes and primers to screen normal and tumorous tissues for the expression of individual CEA-related genes. These and other studies have shown that, in general, NCA mRNA levels are significantly higher in colon adenomas and adenocarcinomas than in normal colonic mucosa, whereas CEA mRNA levels do not change dramatically upon malignant transformation (Boucher et al. 1989; Sato et al. 1988; Cournoyer et al. 1988; Higashide et al. 1990; Hinoda et al. 1991). In order to be able to study large numbers of tissue samples we have developed an assay system where we can specifically identify CEA, NCA, BGP, CGM1 and CGM6 mRNAs using the polymerase chain reaction (PCR). The feasibility of this approach is currently being tested with a larger number of gynecological tumors. RNAs, the integrity of which has been proven by amplification of a β-actin mRNA fragment, are reacted with a pair of primers recognizing all known members of the CEA gene family. The positive samples are then analyzed for the presence of the mRNA of each of the above-mentioned members of the CEA gene family. Preliminary results indicate that, in general, CEA mRNA is coexpressed with NCA mRNA in tumors. Therefore, the CEA and NCA genes, which are next to each other in the CEA gene locus, might share common regulatory elements. Interestingly, most mucinous ovarian carcinomas contained both CEA and NCA mRNAs, whereas ovarian adenocarcinomas of the serous subtype did not express any of the CEA-related mRNAs tested. The expression pattern of the CEA gene family at the mRNA level is summarized in Table 3.

Clinical Implications

What is the relevance of these findings for the diagnosis and therapy of benign and malignant disease? The sensitivity of detection and targeting of CEA could possibly be increased by using cocktails of CEA-specific antibodies. Since we have demonstrated that most of the so far characterized members of the CEA gene subgroup (with the exception of the CEA gene) are expressed on granulocytes, the commonly used approach to test anti-

Table 3. Expression pattern of the CEA gene subgroup

Gene	mRNA size (kb)	Encoded protein	Tissue or cells
CEA	3.5, 3.0	CEA	Normal colon mucosa Colonic polyps Colonic adenocarcinomas (~100%) Less in other carcinomas of epithelial origin (mucinous ovarian carcinomas, lung, pancreas)
BGP	3,9, 3.7, 2.2, 1.8	BGP I (NCA-160)	Normal hepatocyte Hepatocellular carcinoma
NCA	2.5	NCA-50/90	Seems to be always coexpressed with CEA, though to a lesser degree in normal colonic mucosa and polyps CML leukocytes Bone marrow
CGM1	1.3	?	CML leukocytes
CGM2	?	?	?
CGM6	2.2	NCA-95	CML leukocytes Bone marrow

mRNA levels were assessed by northern blot or PCR analyses.

CEA antibodies for cross-reaction against granulocytes is still appropriate, if cross-reactivity with PSGs can be excluded. The monitoring of patients with colorectal tumors might be improved by measuring NCA serum concentrations. Preliminary findings from analysis of a large number of sera of tumor patients, patients with benign disease and healthy individuals with certain antibody combinations which recognize single family members or sets of CEA-related antigens indicate that the sensitivity of tumor detection can be increased without unacceptable loss of specificity. Furthermore, NCA could turn out to be useful as a marker for tumor progression, as the expression of the NCA gene seems to increase with progressing malignancy. The PCR technology would allow determination of NCA mRNA levels in small amounts of biopsy material. Radiolabeled antibodies specific for the CGM6 product might improve the detection of occult inflammatory lesions. Presently, for this purpose, CEA cross-reactive antibodies are employed (D'Amico et al. 1991). CEA and NCA mRNAs (and possibly the corresponding proteins) represent biological markers for certain tumor subtypes as shown for ovarian carcinomas. Therefore, identification of these mRNAs or proteins, respectively, might aid diagnosis in tumor cases of ambiguous histology. These promising results, however, have to be confirmed by analyses of a larger number of tumors and sera, as well as corresponding normal tissues.

References

Aurivillius M, Hansen OC, Lazrek MBS, Bock E, Öbrink B (1990) The cell adhesion molecule Cell-CAM 105 is an ecto-ATPase and a member of the immunoglobulin superfamily. FEBS Lett 264:267–269

Benchimol S, Fuks A, Jothy S, Beauchemin N, Shirota K, Stanners CP (1989) Carcinoembryonic antigen, a human tumor marker, functions as an intercellular adhesion molecule. Cell 57:327–334

Berling B, Kolbinger F, Grunert F, Thompson JA, Brombacher F, Buchegger F, von Kleist S, Zimmermann W (1990) Cloning of a carcinoembryonic antigen family member expressed in leukocytes of chronic myeloid leukaemia patients and bone marrow. Cancer Res 50:6534–6539

Bischoff-Delaloye A, Delaloye B, Buchegger F, Gilgien W, Studer A, Curchod S, Givel J-C, Mosimann F, Pettavel J, Mach J-P (1989) Clinical value of immunoscintigraphy in colorectal carcinoma patients: a prospective study. J Nucl Med 30:1646–1656

Bordes M, Knobel S, Martin F (1975) Carcinoembryonic antigen (CEA) and related antigens in blood cells and haematopoietic tissues. Eur J Cancer 11:783–786

Boucher D, Cournoyer D, Stanners CP, Fuks A (1989) Studies on the control of gene expression of the carcinoembryonic antigen family in human tissue. Cancer Res 49:847–852

Branscomb E, Slezak T, Pae R, Galas D, Carrano AV, Waterman M (1990) Optimizing restriction fragment fingerprinting methods for ordering large genomic libraries. Genomics 8:351–366

Buchegger F, Vacca A, Carrel S, Schreyer M, Mach J-P (1988) Radioimmunotherapy of human colon carcinoma by [131I]-labelled monoclonal anti-CEA antibodies in a nude mouse model. Int J Cancer 41:127–134

Cournoyer D, Beauchemin N, Boucher D, Benchimol S, Fuks A, Stanners CP (1988) Transcription of genes of the carcinoembryonic antigen family in malignant and nonmalignant human tissues. Cancer Res 48:3153–3157

D'Amico P, Lastoria S, Caccavella N, Salvatore M (1991) Radiolabelled granulocytes in inflammatory bone disease. Int J Rad Appl Instrum [3] 18:145–147

Fantini GA, DeCosse JJ (1990) Surveillance strategies after resection of carcinoma of the colon and rectum. Surg Gynecol Obstet 171:267–273

Fearon ER, Cho KR, Nigro JM, Kern SE, Simons JW, Ruppert JM, Hamilton SR, Preisinger AC, Thomas G, Kinzler KW, Vogelstein B (1990) Identification of a chromosome 18q gene that is altered in colorectal cancers. Science 247:49–56

Gold P, Freedman SO (1965) Demonstration of tumor-specific antigens in human colonic carcinomata by immunological tolerance and absorption techniques. J Exp Med 121:439–462

Hammarström S, Shively JE, Paxton RJ, Beatty BG, Larsson A, Ghosh R, Börmer O, Buchegger F, Mach J-P, Burtin P, Seguin P, Darbouret B, Degorce F, Sertour J, Jolu JP, Fuks A, Kalthoff H, Schmiegel W, Arndt R, Klöppel G, von Kleist S, Grunert F, Schwarz K, Matsuoka Y, Kuroki M (1989) Antigenic sites in carcinoembryonic antigen. Cancer Res 49:4852–4858

Higashide T, Hinoda Y, Itoh J, Takahashi H, Satoh Y, Ibayashi Y, Imai K, Yachi A (1990) Detection of mRNAs of carcinoembryonic antigen and nonspecific cross-reacting antigen genes in colorectal adenomas and carcinomas by in situ hybridization. Jpn J Cancer Res 81:1149–1154

Hinoda Y, Imai K, Nakagawa N, Ibayashi Y, Nakano T, Paxton RJ, Shively JE, Yachi A (1990) Transcription of biliary glycoprotein I in malignant and non-malignant human liver tissues. Int J Cancer 45:875–878

Hinoda Y, Takahashi H, Higashide T, Nakano T, Arimura Y, Yoshimoto M, Imai K, Yachi A (1991) Correlated expression of mRNAs of carcinoembryonic antigen and nonspecific cross-reacting antigen genes in malignant and nonmalignant tissues of the colon. Jpn J Clin Oncol 21:75–81

Kuhn JA, Beatty BG, Wong JY, Esteban JM, Wanek PM, Wall F, Buras RR, Williams LE, Beatty JD (1991) Interferon enhancement of radioimmunotherapy for colon carcinoma. Cancer Res 51:2335–2339

Leusch H-G, Hefta SA, Drzeniek Z, Hummel K, Markos-Pusztai Z, Wagener C (1990) Escherichia coli of human origin binds to carcinoembryonic antigen (CEA) and non-specific crossreacting antigen (NCA). FEBS Lett 261:405–409

Leusch H-G, Drzeniek Z, Markos-Pusztai Z (1991) Binding of Escherichia coli and Salmonella strains to member of the carcinoembryonic antigen family: differential binding inhibition by aromatic α-glycosides of mannose. Infect Immun 59:2051–2057

Lin S-H, Guidotti G (1989) Cloning and expression of a cDNA coding for a rat liver plasma membrane ecto-ATPase. J Biol Chem 264:14408–14414

Öbrink B (1991) C-CAM (Cell-CAM 105) – a member of the growing immunoglobulin superfamily of cell adhesion proteins. Bioessays 13:227–234

Oikawa S, Inuzuka C, Kuroki M, Matsuoka Y, Kosaki G, Nakazato H (1989) Cell adhesion activity of non-specific cross-reacting antigen (NCA) and carcinoembryonic antigen (CEA) expressed on CHO cell surface: homophilic and heterophilic adhesion. Biochem Biophys Res Commun 164:39–45

Oikawa S, Inuzuka C, Kuroki M, Arakawa F, Matsuoka Y, Kosaki G, Nakazato H (1991) A specific heterotypic cell adhesion activity between members of carcinoembryonic antigen family, W272 and NCA, is mediated by N-domains. J Biol Chem 266:7995–8001

Pelgrin A, Folli S, Buchegger F, Mach JP, Wagnieres G, van den Bergh H (1991) Antibody-fluorescein conjugates for photoimmunodiagnosis of human colon carcinoma in nude mice. Cancer 67:2529–2537

Pignatelli M, Durbin H, Bodmer WF (1990) Carcinoembryonic antigen functions as an accessory adhesion molecule mediating colon epithelial cell-collagen interactions. Proc Natl Acad Sci USA 87:1541–1545

Rojas M, Fuks A, Stanners CP (1990) Biliary glycoprotein (BGP), a member of the immunoglobulin supergene family, functions in vitro as a Ca^{++}-dependent intercellular adhesion molecule. Cell Growth Differ 1:527–533

Sato C, Miyaki M, Oikawa S, Nakazato H, Kosaki G (1988) Differential expression of carcinoembryonic antigen and nonspecific crossreacting antigen genes in human colon adenocarcinomas and normal colon mucosa. Jpn J Cancer Res 79:433–437

Sharkey RM, Weadock KS, Natale A, Haywood L, Aninipot R, Blumenthal RD, Goldenberg DM (1991) Successful radioimmunotherapy for lung metastasis of human colonic cancer in nude mice. J Natl Cancer Inst 83:627–632

Shively JE, Beatty JD (1985) CEA-related antigens: molecular biology and clinical significance. Crit Rev Oncol Hematol 2:355–399

Svenberg T (1976) Carcinoembryonic antigen-like substances of human bile: isolation and partial characterization. Int J Cancer 17:588–596

Thompson J, Zimmermann W (1988) The carcinoembryonic antigen gene family: structure , expression and evolution. Tumour Biol 9:63–83

Thompson J, Zimmermann W, Osthus-Bugat P, Schleussner C, Eades-Perner A-M, Barnet S, von Kleist S, Willcocks T, Craig I, Tynan K, Olsen A, Mohrenweiser H (1992) Long-range chromosomal mapping of the carcinoembryonic antigen (CEA) gene family cluster. Genomics 12:761–772

Thompson JA, Grunert F, Zimmermann W (1991) The carcinoembryonic antigen gene family: molecular biology and clinical perspectives. J Clin Lab Anal 5:344–366

Turbide C, Rojas M, Stanners CP, Beauchemin N (1991) A mouse carcinoembryonic antigen (CEA) gene family member is a calcium dependent cell adhesion molecule. J Biol Chem 266:309–315

von Kleist S, Burtin P (1966) Mise en évidence dans les tumeurs coliques humaines d'ántigènes non présents dans la muqueuse colique de l'adulte normal. C R Acad Sci (Paris) 263:1543–1546

Watt SM, Sala-Newby G, Hoang T, Gilmore DJ, Grunert F, Nagel G, Murdoch SJ, Tchilian E, Lennox ES, Waldmann H (1991) CD 66 identifies a neutrophil-specific epitope within the hematopoietic system that is expressed by members of the carcinoembryonic antigen family of adhesion molecules. Blood 78:63–74

Williams AF (1987) A year in the life of the immunoglobulin superfamily. Immunol Today 8:298–303

Williams RK, Jiang G-S, Holmes KV (1991) Receptor for mouse hepatitis virus is a member of the carcinoembryonic antigen family of glycoproteins. Proc Natl Acad Sci USA 88:5533–5536

Zhou H, Fuks A, Stanners CP (1990) Specificity of intercellular adhesion mediated by various members of the immunoglobulin supergene family. Cell Growth Differ 1:209–215

Membrane Proteins as Markers for Normal and Neoplastic Endocrine Cells

Georgia Lahr and Manfred Gratzl

Abteilung Anatomie und Zellbiologie, Universität Ulm, Albert-Einstein-Allee 11, W-7900 Ulm, FRG

Introduction

Neurons, endocrine cells and their neoplastic derivatives share a variety of similar or even identical characteristic proteins. The analysis of these cellular constituents provides valuable information on the nature, location and distribution of tumor cells in the body.

Neuron-specific enolase, a soluble cytoplasmic glycolytic enzyme, was the first widely used marker protein for neural and endocrine cells and serves as an excellent diagnostic tool in neurologic and endocrine disease [22]. Recently, specific secretory products of endocrine cells, the chromogranins/secretogranins, which are costored and coreleased with catecholamines and peptide hormones, have also attracted interest as neural or endocine cell markers. These markers can be measured in tumor tissue or cells and also in serum [15, 36, 37, 41].

In addition to the proteins mentioned above, intracellular membrane proteins or constituents of the plasma membrane have very recently been accepted as diagnostic markers. One group of intracellular membrane proteins common to neural and endocrine cells is found in small translucent vesicles (SVs). In neurons SVs contain transmitters and are thus termed synaptic vesicles [6, 16]. In endocrine cells the function of SVs, which contain the same membrane proteins as synaptic vesicles, is unknown. If SVs in endocrine cells also contain transmitters, they could reflect the existence of a second secretory pathway in addition to the well-characterized release of peptide hormones from large, dense core vesicles (LVs) by exocytosis.

Cell surface antigens shared by neural and endocrine cells have also been identified. For example, the neural cell adhesion molecules (NCAMs), have been found in peptide hormone- as well as steroid hormone-producing endocrine cells. Moreover, NCAM expression characterizes the cell-cell adhesion specificities between these cells [18–21, 23, 27]. In addition,

Wagener/Neumann (Eds.)
Molecular Diagnostics of Cancer
© Springer-Verlag Berlin Heidelberg 1993

NCAMs have been found in a variety of rodent and human endocrine tumor cells, suggesting a possible application as markers [1, 2, 9, 10, 13, 17–19, 21, 24, 26, 31].

In this chapter we will depict the structure and function of synaptophysin, as an example of a SV membrane protein already used as an endocrine tumor marker, and describe the molecular and cell biology of NCAMs, a recently elucidated group of membrane antigens present in normal and neoplastic endocrine cells.

Structure and Function of Synaptophysin

Neurotransmission between a presynaptic nerve terminal and postsynaptic target cell involves discharge of transmitter molecules. The neurotransmitters are stored in specialized organelles, the synaptic vesicles. The specific function of synaptic vesicles includes uptake and storage of neurotransmitters, interaction with the cytoskeleton, docking and fusion with the plasma membrane. Synaptophysin [6, 16] is one of the major integral membrane proteins of SVs (average diameter 50 nm). In the SV membrane it forms hexamers, composed of identical subunits of 38 kDa, whose primary structure has been elucidated by sequencing rat, cow and human cDNA. Synaptophysin is a highly conserved protein consisting of four transmembrane domains as well as short N-terminal and extended C-terminal cytoplasmic domains (Fig. 1).

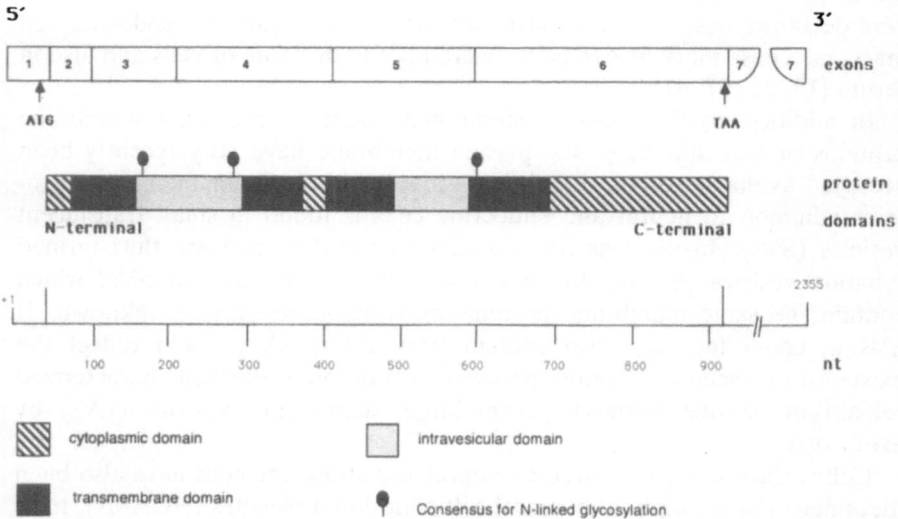

Fig. 1. Schematic representation of the synaptophysin mRNA illustrating the positions of the protein domains coded by the different exons. Exons are represented by the *white boxes*. The start (*ATG*) and stop (*TAA*) codons for translation are also indicated. The other boxes are explained at the foot of the figure. The positions of nucleotides are given in the *lower panel*

Synaptophysin incorporated into lipid membranes exhibits voltage-dependent channel activity [39] and it may dock to a complementary protein, physophilin, a component of the presynaptic plasma membrane, possibly initiating formation of a fusion pore and subsequent neurotransmitter release [38].

Isolation and comparison of the complete rat and human synaptophysin genes showed that, despite the difference in molecular size (16 kb in rat vs 13 kb in man), intron/exon boundaries are precisely conserved [3]. Exon 7, because of a stop codon in exon 6, is not translated in either species (Fig. 1). The 5' upstream region is devoid of any TATA or CAAT boxes but is characterized instead by features typical of "housekeeping" genes. Specifically, GC-rich islands and four Sp1-binding motifs indicate that synaptophysin synthesis is regulated at the transcriptional level. Sequences more than 1.2 kb downstream from the immediate upstream region may be responsible for cell type-specific expression of synaptophysin.

Molecular and Cell Biology of Neural Cell Adhesion Molecules

Cell-cell adhesion molecules (CAMs) – the ligands or receptors involved in cell-cell contact formation – are important regulators of cell assembly and maintenance of tissue architecture.

The NCAMs are among the most prevalent CAMs in vertebrates. They are glycoproteins anchored in different ways to the plasma membrane and promote cell-cell adhesion through a homophilic binding mechanism. The different NCAMs are primary translation products arising through alternative splicing of a single gene located on human chromosome 11 and mouse chromosome 9. They have similar extracellular domains but differ primarily in their plasma membrane associated and intracellular domains. For example, NCAM-180 (the isoform having a molecular weight of 180 kDa) contains a cytoplasmic domain which is larger than that of NCAM-140. By contrast, NCAM-120 consists only of extracellular domains and is linked to the plasma membrane by a phosphatidylinositol anchor (Fig. 2).

Alternative splicing can allow the fine modulation of gene expression such that protein isoforms with functional differences are expressed in the proper spatiotemporal fashion. It is a posttranscriptional process by which pre-mRNA transcribed from the gene can code a variety of protein isoforms that differ in function or localization. Twenty major exons in both mammals and birds code for the different NCAM isoforms. While exon O encodes the 5'-untranslated and leader sequences [33], exons 1–14 generate the extracellular sequences of all known NCAMs (Fig. 2). Exon 15 codes for the membrane-anchoring sequence of NCAM-120, and exon 18 codes for the additional cytoplasmic insert unique to NCAM-180 [4, 5, 25, 32].

Recently, additional exons in the NCAM gene have been discovered. They can be added at the exon 12/13, 13/14 junction, thereby coding for the

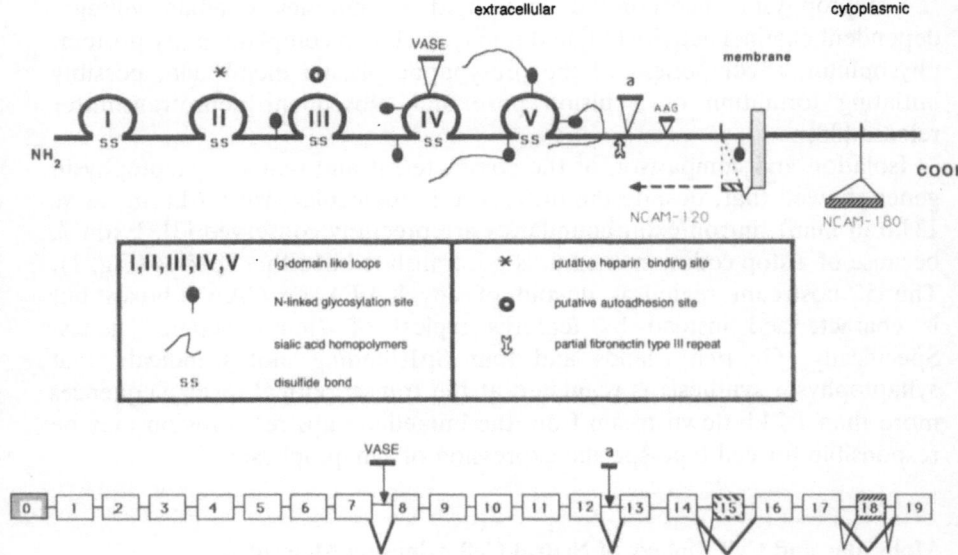

Fig. 2. Scheme of the protein and exon structure of the neural cell adhesion molecule with a molecular weight of 140 kDa (NCAM-140). The N- and C-termini of the protein are indicated. *Loops I–V* represent the disulfide-linked loop structures of the immunoglobulin-like domains. The *inset* explains the symbols used in the upper panel. The plasma membrane is indicated by the *dotted box*. NCAM-120 contains the translation product of exon 15 which is attached to the membrane via a phosphatidylinositol anchor. The sequence coded by exon 18 which is unique to NCAM-180 is shown at the right side of the schema. Minor exons are indicated by *VASE*, *a* and *AAG*. In the *lower panel* the exons of NCAM are represented as boxes. The characteristic exons of NCAM-120 and NCAM-180 are marked in addition to exon 0, which is not translated. The alternatively spliced minor "extra-exons" at exon junctions 7/8 and 12/13 are also shown. The *V* marks indicate alternative splice junctions

hinge region of the molecule, or at the exon 7/8 junction. Genomic cloning revealed that three smaller exons of 15, 48 and 42 bp could be positioned between exons 12 and 13 [8, 28, 40]. Furthermore, an exon named SEC, which contains a stop codon generating a secreted NCAM form, has been spliced into this position [11]. A 15-nt exon (a15) with or without an additional AAG triplet has also been detected between exon 12 and exon 13 [33]. It is not clear whether the 3'-terminal AAG triplet is encoded by a 3-nt exon [33]. An AAG triplet has also been found at the exon 13/14 junction (D. Barthels et al. 1991, personal communication). From the examples given above it is obvious that the hinge region of the NCAMs can be modified by alternative splicing in many ways. The functional consequences of these insertions are as yet unknown. The structure of the N-terminal extracellular regions of NCAMs contains five domains which are similar to each other and to the homology units of immunoglobulins.

Thus NCAMs, like other cell adhesion molecules such as L1, F1, contactin, TAG-1 and MAG, are members of the immunoglobulin superfamily.

A heparin-binding domain has been identified in the second immunoglobulin-like domain of NCAMs. Additionally, an N-linked oligo-saccharide has been mapped to the third immunoglobulin-like domain which is engaged in the autoadhesion between NCAMs of neighboring cells. It is the adjacent fourth immunoglobulin-like loop of NCAMs which can be varied by alternative splicing. A 30-nt exon, termed pi or VASE (variable alternatively spliced exon), is located between exons 7 and 8 [33, 35]; it has been suggested that VASE may modify binding affinity [34]. Taken together, 198 different mRNA species could code for NCAM polypeptides (D. Barthels et al. 1991, personal communication) [30, 33]. The presence of extra exons in cellular mRNA can be detected using S1 nuclease protection assays. This method allows direct comparison of mRNA and labeled DNA or RNA probe sequences. The S1 nuclease hydrolyzes only single-stranded DNA and RNA probes. Thus, if the DNA or RNA probe does not pair precisely, tails or loops not protected by hybridization to the mRNA will be excised. Resolution of the undigested S1-protected fragments of the probes on denaturing polyacrylamide gels yields detailed information about the regions of sequence homology between the probe and the mRNA (Fig. 3).

The fifth extracellular loop of NCAMs is subject to functionally important posttranslational nodification. Though NCAMs have multiple carbohydrate attachment sites, modulation of adhesion specifically arises from differences in the length of homopolymers of alpha-2, 8-linked neuraminic acid units (polysialic acid, PSA) linked to NCAMs via a core carbohydrate. NCAMs occur in highly sialylated embryonic and less sialylated adult forms. The precise molecular mechanism whereby polysialic acid on NCAMs modulates the calcium-independent autoadhesion remains to be analyzed, but it is clear that enzymatic removal of PSA increases binding between NCAM-bearing liposomes and neuroblastoma cells [29].

Following transient expression in various embryonic structures, NCAMs are subsequently found predominantly in adult neural, endocrine, skeletal and cardiac muscle cells. Available data imply that NCAMs influence a number of developmental processes such as segregation of cells into discrete regions, axon guidance and formation and innervation of skeletal muscle [4, 5, 23, 25, 32].

Regulation of gene expression can occur at the transcriptional, post-trancriptional, translational and posttranslational levels as developmental processes unfold. Regulation at the transcription level is initiated by the promoter region of the gene. It should be noted that the NCAM promoter does not contain a typical TATA box, and therefore initiation of RNA transcription occurs at several sites on the gene; this initiation mechanism is often observed in genes that lack this sequence. Sequences for both promotion and inhibition of transcription reside within 840 bp upstream of the main transcription start site. In this upstream region a juxtaposition

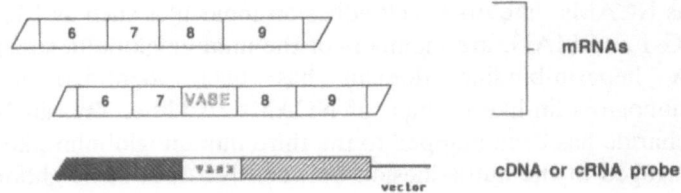

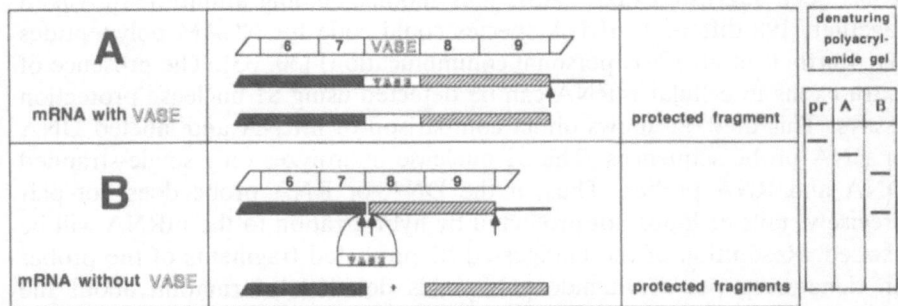

Fig. 3. Detection of additional exons in NCAM mRNA by S1 nuclease protection assays. The exons of two different mRNAs as well as the labeled cDNA or cRNA probe with additional vector sequences used are shown at the top. The probe is hybridized with the mRNAs with (*A*) or without (*B*) exon VASE followed by S1 nuclease digestion, which hydrolyzes all single-stranded regions (indicated by the *arrows*). In *A* only the vector sequences are removed, whereas in *B* the single strand coding for extra-exon VASE contained in the labeled probe will be totally digested by S1 nuclease, leaving two protected fragments. The samples are run on a denaturing polyacrylamide gel to determine the size of probe fragments protected from the S1 nuclease by hybridization

of a Sp1 factor-binding consensus site and a nuclear factor I-binding site has been mapped. Moreover, motifs, such as three A + T-rich segments containing ATTA motifs and an AGGA repeat which resembles negative regulatory elements in other promoters, were also mapped. It seems that negative and positive elements in the promoter, with features typical of "housekeeping" genes, interact to regulate the tissue-specific pattern of NCAM gene expression. Furthermore, a factor related to nuclear factor I is involved in transcriptional control of the NCAM gene [14].

Expression of Synaptophysin and NCAMs in Normal and Neoplastic Endocrine Cells

The distribution of synaptophysin and NCAMs in normal and neoplastic endocrine cells is summarized in Table 1. In addition to their occurrence in

Table 1. Occurrence of the antigen of small secretory vesicles, synaptophysin (SYN), and of the neural cell adhesion molecules (NCAMs) in normal and neoplastic endocrine cells and tissues

Tissues	Cells and tumors	SYN	NCAMs
Adenohypophysis	Normal endocrine cells	+	+
	Adenoma	+	+
Adrenal medulla	Normal chromaffin cells	+	+
	Pheochromocytoma	+	+
Adrenal cortex	Normal steroid-producing cells	−	+
	Cortical carcinoma	n.d.	+
Bronchial tract	Normal endocrine cells	+	n.d.
	Small cell carcinoma	+	+
	Carcinoid	+	n.d.
Gastrointestinal tract	Normal enteroendocrine cells	+	n.d.
	Carcinoid	+	n.d.
Pancreas	Normal islet cells	+	+
	Adenoma	+	n.d.
Parathyroid	Normal chief cells	+	n.d.
	Adenoma	+	n.d.
Thyroid	Normal C cells	+	n.d.
	C-cell carcinoma	+	n.d.
Testis	Leydig cells	n.d.	+
Ovary	Granulosa and luteal cells	n.d.	+

n.d., not determined.

catecholamine- and peptide hormone-containing cells, NCAMs have also been detected in steroid-producing cells of adrenal cortex [27] and ovarian granulosa and luteal cells [23]. The potential application of NCAM analysis in tumors derived from these tissues can be envisioned, based on a report of NCAM expression in an adrenocortical carcinoma [17]. Synaptophysin, on the other hand, occurs in all normal peptidergic endocrine cells and in neoplasms derived from these cells (Table 1).

A variety of different techniques has been applied in the analysis of synaptophysin and NCAMs in normal and neoplastic endocrine cells. These analytical techniques provide precise descriptions of the extent of transcription, translation and even posttranslational modification of cellular membrane proteins. For example, transciption is analysed using in situ hybridization and Northern blot techniques in addition to the more sophisticated S1 nuclease protection assays, described above, which can elucidate the sequence of a specific mRNA. Insights into translation of synaptophysin and NCAMs in particular cells are provided by light and ultrastructural immunocytochemistry. These morphological techniques, in conjunction with immunological techniques such as Western blotting, can be utilized to analyze the anatomical arrangements related to posttranslational processing or modification (e.g. proteolytic modification, presence of PSA and others). Taken together, these techniques have been used to define the nature of certain primary tumors and, importantly, to detect their metastases. It is also possible to identify neoplastic cells which express endocrine marker proteins in mixed tumors. For example, cell heterogeneity is a prominent characteristic in most lung neoplasms. Yet, in mixed tumors as well as in homogeneous tumors immunocytochemical techniques reveal different cell types based on their expression of NCAM-140; specifically, small cell carcinoma cells express NCAM-140 while mesotheliomas and squamous cell carcinomas do not [2, 9, 21, 24, 26]. Differential diagnosis of the lung tumors based on their expression of cell markers provides important insights for prognosis and therapy.

Detection of synaptophysin (like neuron-specific enolase) can certainly be regarded as a general tool to distinguish endocrine from nonendocrine tumor cells. NCAMs share with other endocrine markers the property of differential expression in endocrine tumor cells. For example, all pituitary adenomas, characterized by hormone markers [2], have been found to contain synaptophysin and neuron-specific enolase, but NCAM-140, the prevalent isoform in normal and neoplastic endocrine tumor cells, was not detected by immunoblotting in prolactinomas, suggesting that it is minimally expressed in this cell type [2]. In like manner, chromogranin A, a protein costored and coreleased from a variety of peptide hormone-producing cells, has not been observed in all their neoplasms. While prolactinomas do not appear to contain chromogranin A [7], they can be identified by their hormone content and the expression of synaptophysin and neuron-specific enolase. Adenomas not expressing hormones, termed inactive adenomas, generally express the latter markers together with NCAM-140 and chromogranin A; on the other hand, analysis of the hormones in serum or by immunocytochemistry in these adenomas is not of diagnostic value [2, 7, 12].

The examples given above clearly demonstrate that analysis of membrane markers for endocrine tumor cells provides important contributions to a precise (differential) diagnosis and elucidation of the biological properties of the neoplastic cells even though they differ in their degrees of proliferation and their tendencies to form metastases. Thus, membrane marker characterizations form an important basis for therapeutic design and modulation of its effect.

Acknowledgments. Studies from the authors' laboratories were supported by the Deutsche Krebshilfe, Deutsche Forschungsgemeinschaft and Landesforschungsschwerpunkt 32 of Baden-Württemberg. The authors thank B. Mader for preparation of the manuscript.

References

1. Aletsee-Ufrecht MC, Langley OK, Gratzl O, Gratzl M (1990) Differential expression of the neural cell adhesion molecule NCAM140 in human pituitary tumors. FEBS Lett 272:45–49
2. Aletsee-Ufrecht MC, Langley OK, Rotsch M, Havemann K, Gratzl M (1990) NCAM: a surface marker for human small cell lung cancer cells. FEBS Lett 67:295–300
3. Bargou RCEF, Leube RE (1991) The synaptophysin-encoding gene in rat and man is specifically transcribed in neuroendocrine cells. Gene 99:197–204
4. Barthels D, Vopper G, Wille W (1988) NCAM-180, the large isoform of the neural cell adhesion molecule of the mouse, is coded by alternatively spliced transcript. Nucleic Acids Res 16:4217–4225
5. Cunningham BA, Hemperly JJ, Murray BA, Prediger EA, Bruckenbury R, Edelman GM (1987) Neural cell adhesion molecule: structure, immunoglobulin-like domains, cell surface modulation, and alternative RNA splicing. Science 236:799–805
6. de Camilli P, Jahn R (1990) Pathways to regulated exocytosis in neurons. Annu Rev Physiol 52:625–645
7. Deftos LJ (1991) Chromogranin A: its role in endocrine function and as an endocrine and neuroendocrine tumor marker. Endocr Rev 12:181–187
8. Dickson GH, Gower HJ, Barton CH, Prentice HM, Elsom VL, Moore SE, Cox RD, Quinn C, Putt W, Walsh FS (1987) Human muscle cell adhesion molecule (N-CAM): identification of a muscle-specific sequence in the extracellular domain. Cell 50:1119–1130
9. Doyle LA, Borges M, Hussain A, Elias A, Tomiyasu T (1990) An adherent subline of a unique small-cell lung cancer cell line downregulates antigens of the neural cell adhesion molecule. J Clin Invest 86:1848–1854
10. Figarella-Branger DF, Durbec PL, Rougon GN (1990) Differential spectrum of expression of neural cell adhesion molecule isoforms and L1 adhesion molecules on human neuroectodermal tumors. Cancer Res 50:6384–6370
11. Gower HJ, Barton CH, Elsom VM, Tompson J, Moore SE, Dickson JG, Walsh FS (1988) Alternative splicing generates a secreted form of N-CAM in muscle and brain. Cell 55:955–964
12. Heitz PU, Landolt AM, Zenklusen H-R, Kasper M, Reubi J-C, Oberholzer M, Roth J (1987) Immunocytochemistry of pituitary tumors. J Histochem Cytochem 35:1005–1011
13. Heitz PU, Roth J, Zuber C, Komminoth P (1991) Markers for neural and endocrine cells in pathology. In: Gratzl M, Langley K (eds) Markers for neural and endocrine cells. VCH, Weinheim, pp 203–216
14. Hirsch M-R, Gaugler L, Deagostini-Bazin H, Bally-Cui L, Goridis C (1990) Identification of positive and negative regulatory elements governing cell-type-specific expression of the neural cell adhesion molecule gene. Mol Cell Biol 10:1959–1968
15. Huttner WB, Gerdes H-H, Rosa P (1991) Chromogranins/secretogranins – widespread constituents of the secretory granule matrix in endocrine cells and neurons. In: Gratzl M, Langley K (eds) Markers for neural and endocrine cells. VCH, Weinheim, pp 93–131
16. Jahn R, de Camilli P (1991) Membrane proteins of synaptic vesicles: markers for neurons and neuroendocrine cells, tools for the study of neurosecretion. In: Gratzl M, Langley K (eds) Markers for neural and endocrine cells. VCH, Weinheim, pp 25–92

17. Jin L, Hemperly JJ, Lloyd RV (1991) Expression of neural cell adhesion molecule in normal and neoplastic human neuroendocrine tissues. Am J Pathol 138:961–969
18. Langley OK, Aletsee MC, Gratzl M (1987) Endocrine cells share expression of N-CAM with neurones. FEBS Lett 220:108–112
19. Langley OK, Aletsee-Ufrecht MC, Grant NJ, Gratzl M (1989) Expression of the neural cell adhesion molecule NCAM in endocrine cells. J Histochem Cytochem 37:781–791
20. Langley OK, Aunis D (1984) Ultrastructural immunocytochemical demostration of D2-protein in adrenal medulla. Cell Tissue Res 238:497–502
21. Langley OK, Gratzl M (1991) Neural cell adhesion molecule NCAM in neural and endocrine cells. In: Gratzl M, Langley K (eds) Markers for neural and endocrine cells. VCH, Weinheim, pp 133–178
22. Marangos PJ (1991) Neuron specific enolase as a clinical tool in neurologic and endocrine disease. In: Gratzl M, Langley K (eds) Markers for neural and endocrine cells. VCH, Weinheim, pp 181–189
23. Mayerhofer A, Lahr G, Gratzl M (1991) Expression of the neural cell adhesion molecule (NCAM) in endocrine cells of the ovary. Endocrinology 129:792–800
24. Moolenaar CECK, Muller EJ, Schol DJ, Figdor CG, Bock E, Bitter-Suermann D, Michalides RJAM (1990) Expression of neural cell adhesion molecule-related sialoglycoprotein in small cell lung cancer and neuroblastoma cell lines H69 and CHP-212. Cancer Res 50:1102–1106
25. Nybroe O, Linnemann D, Bock E (1988) NCAM biosynthesis in brain. Neurochem Int 12:251–262
26. Patel K, Moore SE, Dickson G, Rossell RJ, Beverly PC, Kemshead JT, Walsh FS (1989) Neural cell adhesion molecule (NCAM) is the antigen recognized by monoclonal antibodies of similar specificity in small-cell lung carcinoma and neuroblastoma. Int J Cancer 44:573–578
27. Poltorak M, Shimoda K, Freed WJ (1990) Cell adhesion molecules (CAMs) in adrenal medulla in situ and in vitro: enhancement of chromaffin cell L1/Ng-CAM expression by NGF. Exp Neurol 110:52–72
28. Predinger EA, Hoffman S, Edelman GM, Cunningham BA (1988) Four exons encode a 93-base-pair insert in three neural cell adhesion molecule mRNAs specific for chicken heart and skeletal muscle. Proc Natl Acad Sci USA 85: 9616–9620
29. Regan CM (1991) Regulation of neural cell adhesion molecule sialylation state. Int J Biochem 23:513–523
30. Reyes AA, Small SJ, Akeson R (1991) At least 27 alternatively spliced forms of the neural cell adhesion molecule mRNA are expressed during rat heart development. Mol Cell Biol 11:1654–1661
31. Roth J, Zuber C, Wagner P, Taatjes DJ, Weisgerber C, Heitz PU, Goridis C, Bitter-Suermann D (1988) Reexpression of poly (sialic acid) units of the neural cell adhesion molecule in Wilms tumor. Proc Natl Acad Sci USA 85:2999–3003
32. Rutishauser U, Jessel TM (1988) Cell adhesion molecules in vertebrate neural development. Physiol Rev 68:819–857
33. Santoni M-J, Barthels D, Vopper G, Boned A, Goridis C, Wille W (1989) Differential exon usage involving an unusual splicing mechanism generates at least eight types of NCAM cDNA in mouse brain. EMBO J 8:385–392
34. Small SJ, Akeson R (1990) Expression of the unique NCAM VASE exon is independently regulated in distinct tissues during development. J Cell Biol 111:2089–2096
35. Small SJ, Haines SL, Akeson R (1988) Polypeptide variation in N-CAM extracellular immunoglobulin like fold is developmentally regulated through alternative splicing. Neuron 1:1007–1017

36. Takiyyuddin MA, Barbosa JA, Hsiao RJ, Parmer RJ, O'Connor DT (1991) Diagnostic value of chromogranin A measured in the circulation. In: Gratzl M, Langley K (eds) Markers for neural and endocrine cells. VCH, Weinheim, pp 191–201
37. Takiyyuddin MA, Cervenka JH, Pandian MR, Stuenkel CA, Neumann HPH, O'Connor DT (1990) Neuroendocrine sources of chromogranin-A in normal man: clues from selective stimulation of endocrine glands. J Clin Endocrinol Metab 71:360–369
38. Thomas L, Betz H (1990) Synaptophysin binds to physophilin, a putative synaptic plasma membrane protein. J Cell Biol 111:2041–2052
39. Thomas L, Hartung K, Langosch D, Rehm H, Bamberg E, Franke WW, Betz H (1988) Identification of synaptophysin as a hexameric channel protein of the synaptic vesicle membrane. Science 242:1050–1053
40. Thompson J, Dickson G, Moore SE, Gower HJ, Putt W, Kenimer JG, Barton CH, Walsh FS (1989) Alternative splicing of the neural cell adhesion molecule gene generates variant extracellular domain structure in the skeletal muscle and brain. Genes Dev 3:348–357
41. Wiedenmann B, Huttner WB (1989) Synaptophysin and chromogranins/ secretogranins – widespread constituents of distinct types of neuroendocrine vesicles and new tools in tumor diagnosis. Virchows Arch [B] 58:95–121

Biological and Clinical Relevance of the Tumor-Associated Serine Protease uPA

M. Schmitt, F. Jänicke, N. Chucholowski, L. Goretzki, N. Moniwa, E. Schüren, O. Wilhelm, and H. Graeff

Frauenklinik, Klinikum rechts der Isar, Technische Universität München, Ismaningerstraße 22, W-8000 München 80, FRG

Introduction

Two types of serine proteases, uPA (urokinase-type plasminogen activator) and tPA (tissue-type plasminogen activator), are known to convert plasminogen into plasmin [1]. tPA is mainly involved in intravascular thrombolysis [1, 2] whereas uPA mediates pericellular proteolysis during cell migration and tissue remodeling under physiological and pathophysiological conditions [2, 3]. Although initially secreted in the form of an enzymatically inactive, single-chain proenzyme (pro-uPA), uPA exerts its proteolytic function on normal cells and tumor cells as an ectoenzyme after having bound to a specific high-affinity cell surface receptor (uPA-R) [4-6]. uPA-R on cells seems to be the reaction site for uPA-mediated plasminogen activation, also in solid tumors. Quantitative assessments of tumor tissues, invasion assays with tumor explants in experimental animals and in vitro investigations with tumor cells suggest that tumor invasion and metastasis are correlated with elevated levels of uPA and the presence of the uPA-R [2-6]. Tissues of primary cancer and/or metastases of the breast, ovary, prostate, cervix uteri, bladder, lung and gastrointestinal tract have been reported to contain high amounts of uPA compared to benign control tissues [2, 3]. uPA is produced and secreted as pro-uPA by normal cells and by tumor cells [2, 3]. Pro-uPA may be converted by small amounts of serine proteases (plasmin, kallikrein, trypsin) or cathepsin B or L into the enzymatically active, high-molecular-weight, two-chain form HMW-uPA which subsequently converts plasminogen into the serine protease plasmin [7-10]. tPA, which has also been identified in normal cells and in tumor tissues, is not involved in tumor invasion and metastasis [2, 3, 11]. The enzymatic activity of uPA and tPA can be blocked by specific inhibitors, the plasminogen activator inhibitors PAI-1 and PAI-2. uPA, tPA, PAI-1, PAI-2 and uPA-R have been characterized chemically and cloned [1, 2, 12]. Both inhibitors are members of the serpin superfamily of proteins and react rapidly with uPA or tPA. Protease nexin (PAI-3) and protein C inhibitor may also inhibit uPA and tPA activity [1, 13]. Some of the physicochemical characteristics of uPA, tPA, PAI-1 and PAI-2 are listed below.

Wagener/Neumann (Eds.)
Molecular Diagnostics of Cancer
© Springer-Verlag Berlin Heidelberg 1993

uPA. Glycosylated at Asn[302]. Produced by kidney tubule cells, phagocytic cells, pneumocytes, keratinocytes, fibroblasts and tumor cells. Released as pro-uPA. The gene is located on chromosome 10. Gene length 6.4 kb. Several forms of uPA are known. *Pro-uPA*: Single-chain proenzyme form of uPA with 411 amino acids. M_r 52 000. Low enzymatic activity (0.4% of HMW-uPA). Does not bind to PAI-1 and PAI-2. Binds to uPA-R on cell surface membrane. Degraded by serine proteases or cathepsin B or L into HMW-uPA. *HMW-uPA*: 411 amino acids. M_r 52 000. Two chains (A and B) linked by a disulfide bond. A-chain 158 amino acids, B-chain 253 amino acids. Enzymatically active. Binds to PAI-1 and PAI-2. Binds to uPA-R on cell surface membrane. Degraded by serine proteases into the low-molecular-weight two-chain form (LMW-uPA) and the amino-terminal fragment (ATF). *LMW-uPA*: 276 amino acids. M_r 34 000. Two chains (A' and B) linked by disulfide bonds. A'-chain 23 amino acids, B-chain 253 amino acids. Enzymatically active. Binds to PAI-1 and PAI-2. Does not bind to uPA-R. Active center amino acids of HMW-uPA and LMW-uPA: His[204], Asp[255], Ser[356]. *ATF*: 135 amino acids. Amino-terminal part of the A-chain of uPA. M_r 16 000. Consists of the Kringle domain and the growth-factor-like domain (GFD). Enzymatically inactive. Does not bind to PAI-1 and PAI-2. Binds to uPA-R on cell surface membrane. Degraded by endoproteinases V8 or Asp-N into Kringle and GFD. *Kringle*: 86 amino acids. M_r 10 000. Enzymatically inactive. Does not bind to uPA-R, PAI-1 or PAI-2. Function unknown. *GFD*: 49 amino acids. M_r 6000. Enzymatically inactive. Does not bind to PAI-1 and PAI-2.

tPA. Glycosylated at Asn[117], Asn[184], Asn[448]. Produced by endothelial cells and tumor cells. Released as pro-tPA. The gene is located on chromosome 8. Gene length 29 kb. Liver cells (hepatocytes) contain a tPA binding site which functions as a clearance receptor for fluid-phase tPA. *Single-chain tPA*: 530 amino acids. M_r 70 000. Enzymatically active. Binds PAI-1 and PAI-2. Degraded by serine proteases into two-chain tPA. *Two-chain tPA*: 530 amino acids. Enzymatically active. Binds also to clearance receptor on liver cells. Binds PAI-1 and PAI-2. A-chain 276 amino acids, B-chain 254 amino acids. Active center amino acids: His[325], Asp[374], Ser[481].

PAI-1. Size: 381 amino acids. Glycoprotein of M_r 50 000. Single chain. Binds to single-chain and two-chain tPA, HMW-uPA and LMW-uPA but not to pro-uPA. Degraded by plasmin and reactive oxygen metabolites. Stabilized by vitronectin. Produced by platelets, endothelial cells, granulosa cells and tumor cells. The gene is located on chromosome 7. The mRNA variants are 2.4 and 3.4 kb long.

PAI-2. Size: 393 amino acids. Nonglycosylated form (M_r 48 000) and glycosylated form (M_r 70 000) known. Single chain. Binds to single-chain and two-chain tPA, HMW-uPA and LMW-uPA but not to pro-uPA. Produced by trophoblast cells, phagocytic cells and tumor cells. The gene is located on chromosome 18. Found in high concentration in the plasma of pregnant women.

The uPA Molecule

The interaction of uPA-R with the uPA molecule has been characterized in detail. Both pro-uPA and HMW-uPA bind to receptors on the surface of tumor cells and normal cells via a defined peptide sequence (13–30) of the GFD of uPA [6, 14]. (Fig. 1, 2). This epitope resides on the ATF of the A-chain of uPA, which is a result of extensive digestion of HMW-uPA

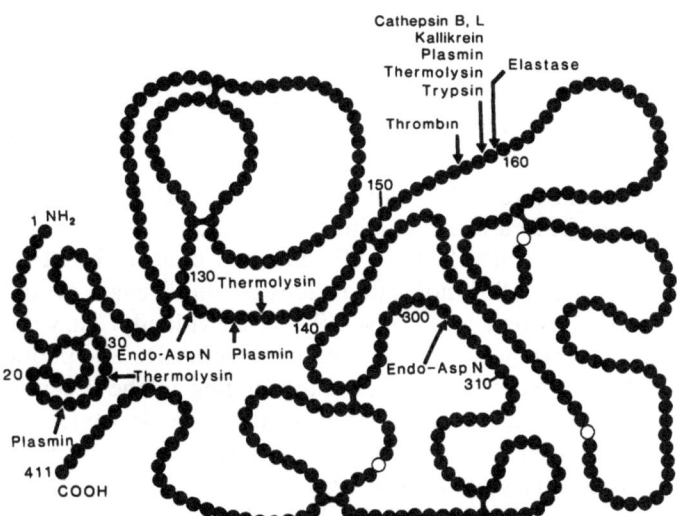

Fig. 1. Cleavage of pro-uPA by proteolytic enzymes yields enzymatically active or degraded HMW-uPA. The proenzyme form of uPA, pro-uPA (M_r 52 000), is synthesized by cells as an enzymatically inactive single-chain molecule consisting of 411 amino acids linked by several disulfide bonds. The serine proteases plasmin, plasma kallikrein and trypsin, as well as thermolysin and cathepsins B and L, cleave pro-uPA at the peptide bond Lys^{158}-Ile^{159}, which converts pro-uPA into the enzymatically active, high-molecular-weight, two-chain form HMW-uPA. This yields two polypeptide chains linked by a disulfide bond: the A-chain (amino acids 1–158, M_r 20 000) and the B-chain (amino acids 159–411, M_r 32 000). Extensive proteolysis of HMW-uPA by plasmin results into additional cleavage of the A-chain of HMW-uPA at peptide bond Lys^{135}-Lys^{136}, which yields the low-molecular-weight, two-chain form LMW-uPA and the amino terminal fragment, ATF. LMW-uPA consists of the B-chain and the residual A-chain linked by disulfide bonds. ATF consists of the Kringle domain and the growth-factor-like domain (GFD). Cleavage of pro-uPA by thrombin at peptide bond Arg^{156}-Phe^{157} or by granulocyte elastase at Ile^{159}-Ile^{160} results in enzymatically inactive HMW-uPA

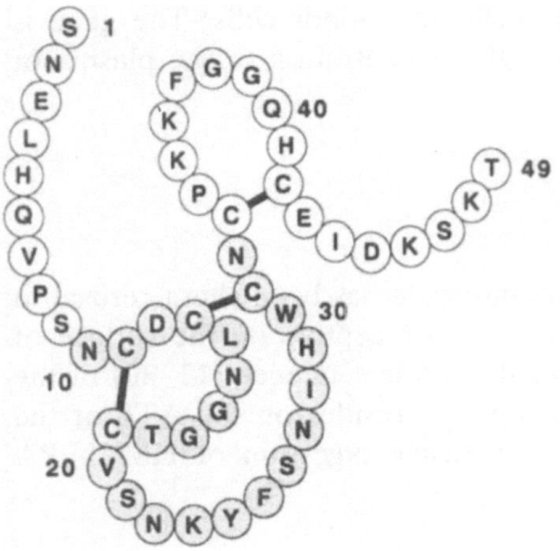

Fig. 2. Receptor-binding domain of human pro-uPA/HMW-uPA. The receptor-binding domain on pro-uPA and also on HMW-uPA resides on a loop within the GFD of ATF. Binding is effected by the peptide sequence 20–30. Amino acids 10–19 are needed to attain proper conformation of GFD. Cleavage of GFD at peptide bond Lys[23]-Tyr[24] destroys the receptor-binding capacity of GFD

by plasmin into ATF and LMW-uPA. HMW-uPA and pro-uPA, but also purified GFD and ATF, bind to the uPA-R with high affinity [6, 14, 15]. LMW-uPA, which lacks ATF, does not bind to the uPA-R. The enzymatic center of the uPA molecule resides on the B-chain of uPA, which is part of both HMW-uPA and LMW-uPA [1]. Evidently, enzymatic activity of uPA is not a prerequisite for binding to the uPA-R. The peptide sequence 13–30 of uPA is necessary for optimum binding to the receptor (Fig. 2). Peptide sequence 20–30 effects binding to the uPA-R, while peptide sequence 13–19 is required to attain the proper conformation of the molecule. Although the secondary structures of EGF (epidermal growth factor) and the GFD of uPA are highly homologous, EGF does not bind to uPA-R. Other closely related molecules, such as human tPA or mouse uPA, also fail to bind to uPA-R [6]. Analysis of mouse and human uPA revealed that there is a difference of seven amino acids in the receptor-binding domain at positions Leu[14], Thr[18], Asn[22], Lys[23], Asn[27], His[29], and Trp[30]. Five of these residues (positions 22, 23, 27, 29 and 30) are also different in the human tPA sequence. The conversion of pro-uPA to enzymatically active HMW-uPA is limited to the specificity of the enzyme applied. Plasmin, plasma kallikrein, trypsin, thermolysin and cathepsin B or L cleave pro-uPA at peptide bond Lys[158]-Ile[159] into enzymatically active HMW-uPA [10]. Proteolytic action of thrombin and granulocyte elastase on pro-uPA, however, results in enzymatically inactive HMW-uPA [8, 9]. Such enzymatically active and inactive uPA forms are indistinguishable by SDS-PAGE [9]. Interestingly, enzymatically active HMW-uPA is not inactivated by subsequent elastase or thrombin action [9].

The uPA-R Molecule

The uPA-R has been described first in 1985 by Vassalli et al. on human monocytes and on the promyelocytic leukemia cell line U937 [16]. It is also present on tumor cell lines derived from solid tumors [17, 18]. The affinity constant for binding of pro-uPA or HMW-uPA to uPA-R is high ($K_a = 10^9 - 10^{10}$ l/m) [6]. Receptor density may vary dramatically: human granulocytes and monocytes have very few uPA-R (between several hundreds and some thousands per cell), while stimulated cultured tumor cells may possess up to a million uPA-R per cell [17–21].

The following features are characteristic for the uPA-R/uPA interaction:

- The binding region of pro-uPA/HMW-uPA is confined to uPA peptide sequence 20–30.
- The affinity constant of binding of pro-uPA to uPA-R is similar to that of HMW-uPA.
- Binding to uPA-R is independent of the enzymatic activity of HMW-uPA.
- Once bound to uPA-R, pro-uPA can be activated by proteases to enzymatically active HMW-uPA.
- Receptor-bound HMW-uPA and pro-uPA are not internalized by the cell. The rate of dissociation is very slow. If, however, receptor-bound HMW-uPA, but not pro-uPA, is inhibited by PAI-1 or PAI-2, rapid internalization of a trimeric complex (uPA-R–HMW-uPA–PAI-1/-2) will occur.

uPA-R was purified from U-937 cells by Nielsen et al. in 1988 [22] and characterized chemically in detail by Behrendt et al. [23] and Ploug et al. [24]. uPA-R is a cysteine-rich glycoprotein of M_r 45 000–55 000. Only 70% of the molecular mass is accounted for by protein; the residual 30% may be released after deglycosylation of uPA-R by N-glycanase (endo-F). The protein sequence deduced from the cDNA of uPA-R would account for a protein consisting of 313 amino acids. The mature uPA-R, however, is only 282 amino acids long due to proteolytic processing in the carboxy-terminal region of nascent uPA-R [25]. The ligand-binding region of uPA-R is confined to the first 87 amino acids [23]. uPA-R is attached to the plasma membrane via a covalent linkage of the carboxy terminus of the protein to a glycosylated form of the phospholipid phosphatidylinositol resulting in a glycolipid anchor termed glycosyl-phosphatidylinositol (GPI) [25]. Functional uPA-R is rapidly released from cells by phosphoinositol-specific phospholipase C by cleaving the glycolipid anchor [25] (Fig. 3). uPA-R has been cloned by Roldan et al. [12].

Assessment of uPA and Its Receptor (uPA-R)

Very sensitive enzymatic and immunologic techniques have been employed to quantify uPA in biological fluids (urine, plasma, ascitic fluid, tissue

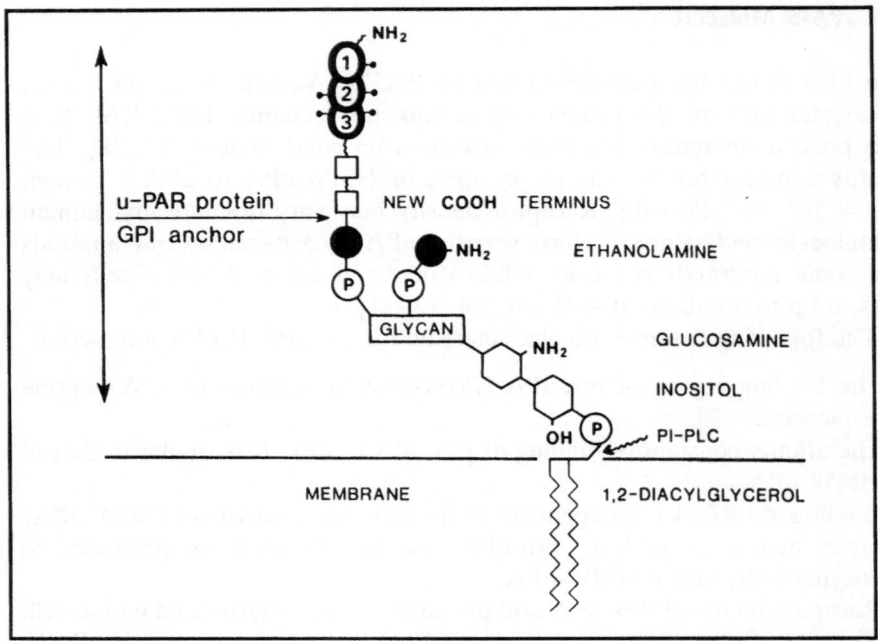

Fig. 3. Domain structure of the uPA-receptor. The uPA-R protein is composed of three internal repeats (domain 1, residues 1–92; domain 2, residues 93–191; domain 3, residues 192–282). The three internal repeats are characterized by a unique pattern of cysteine residues. The uPA-binding epitope resides on domain 1. The basic covalent structure of the glycolipid moiety is largely based on information derived from other GPI-anchored proteins. The linkage between the uPA-R protein and the glycolipid of the plasma membrane occurs via a phosphoethanolamine that forms an amide bond with the alpha-carboxyl group of the uPA-R protein and a phosphodiester bond with the glycan portion of the phospholipid. The diagram was designed by Ploug et al. [24] and is reproduced by permission of Thieme Medical Publishers, New York

extracts) of normal donors and of cancer patients (Table 1). S-2444, a uPA-specific synthetic amidolytic substrate (Pharmacia-Kabi, Stockholm, Sweden, and BACHEM, Bubendorf, Switzerland), has widely been used to assess enzymatic activity of HMW-uPA and LMW-uPA. Sensitive indirect quantitation of uPA activity is also possible by determination of the amount of enzymatically active plasmin generated by the action of uPA or tPA on plasminogen. To demonstrate plasmin activity, the synthetic amidolytic substrate S-2251 (Pharmacia-Kabi) or the fibrin (casein) clot lysis assay (zymography) can be applied.

When using synthetic substrates, measurement of the plasminogen activators uPA and tPA in biological fluids should be done in the presence of quenching antibodies or specific inhibitors [26–28]. Several polyclonal and monoclonal antibodies to uPA and tPA are available commercially (e.g., American Diagnostica, Greenwich, CT, USA). The inhibitory activity

Table 1. Assessment of uPA[a] in cells and in solution

Reaction	Technique(s)	Reference(s)
Reaction of uPA in single cells or tissue cells with: a) Fluorescent antibody to uPA b) Antibody to uPA and then fluoresceinated antibody to immunoglobulin	Fluorescence microscopy, laser scan microscopy, flow cytofluorometry	17, 29, 31, 47, 59
Reaction of uPA in single cells with fluorescent microspheres conjugated with an antibody directed to uPA	Light microscopy, fluorescence microscopy, laser scan microscopy, flow cytofluorometry	48
Reaction of uPA in single cells or tissue cells with antibody to uPA and second enzyme-labeled antibody to immunoglobulin	Immunocytochemistry, immunohistochemistry	41, 49–55
Adsorption of uPA to polymer membrane and then reaction with antibody to uPA (conjugated with enzyme or biotin)	Immunodot	48
SDS-PAGE of uPA, then electrotransfer to polymer membrane and reaction with antibody to uPA (conjugated with enzyme or biotin)	Western blot	7, 57
Reaction with antibody to uPA (catching antibody), then reaction with a second antibody to uPA (conjugated with an enzyme or biotin)	Enzyme-linked immunosorbent assay	41, 51, 74
Competition of binding of radiolabeled uPA to specific antibody by fluid-phase uPA	Radioimmunoassay	16, 21, 58
Enzymatic activity screened with: a) Synthetic substrates b) Conversion of plasminogen to plasmin	Photometry (synthetic substrate) or plaque assay (lysis of fibrin or casein)	7, 9, 48

[a] Pro-uPA, HMW-uPA, LMW-uPA, ATF, Kringle, GFD, synthetic uPA peptides.

of PAI-1 and PAI-2 can be screened by their potential to inhibit uPA or tPA enzymatic activity. Radioimmunoassay (RIA) and enzyme-linked immunosorbent assay (ELISA) techniques have been developed to quantify the uPA, tPA, PAI-1 and PAI-2 antigens. ELISA kits are commercially available (e.g., American Diagnostica, Greenwich, CT, USA; Monozyme,

Copenhagen, Denmark; Biopool, Umeå, Sweden). Indirect immunofluorescence was first applied to visualize receptor-bound uPA antigen on
tumor cells by fluorescence microscopy. uPA, tPA and PAI-1/-2 antigens
can also be detected by immunohistochemistry or immunoelectron microscopy. These immunological techniques can be combined with the "in situ"
hybridization technique to locate cells in tissue sections which do contain
mRNA for these antigens in question.

Several techniques have been employed to assess the uPA-R on normal and on tumor cells (Table 2). Binding of uPA to uPA-R is species-
specific, e.g., mouse uPA does not bind to human uPA-R and vice versa [6].
Initially, radioiodinated HMW-uPA has been used to quantify binding to
suspended or adherent tumor cells. Stoppelli et al. demonstrated binding of
uPA to uPA-R on the adherent epidermoid cell line A431 by the use of a
fluoresceinated antibody directed to uPA [17]. A direct quantification of
receptor-bound uPA on living tumor cells has been described recently in
detail by Schmitt and coworkers [7, 29]. Laser-based flow cytofluorometry
and confocal laser scan microscopy were used by these authors to quantify
binding of FITC-pro-uPA to cell surface uPA-R. Miles et al. [30] and Lu
et al. [31] demonstrated binding of both uPA and plasmin(ogen) to surface
receptors on tumor cells. Monoclonal antibodies (mAbs) to the uPA-R can
be used to locate uPA-R on cells by flow cytofluorometry or by immunohistochemistry [18]. Various synthetic uPA peptides which bind to the uPA-
R have been synthesized. The binding characteristics of such peptides have
been established by Appella et al. [32].

The uPA Antigen Is Elevated in Tumor Tissues

Tumor cell invasion and metastasis are correlated with elevated levels
of proteases such as cathepsins B and D, collagenase IV or plasminogen
activators in tumor tissues [33–39]. uPA, which is secreted by tumor cells as
enzymatically inactive pro-uPA, seems to play a key role in mediating tumor
cell invasion in cancer tissues. For this reason uPA has been the subject
of several detailed biochemical and clinical investigations in recent years
(Table 3).

As early as 1976 Astedt and Holmberg documented the identity of tumor
cell uPA in ovarian cancer ascitic fluid and uPA found in urine [40].
Subsequently, elevated levels of enzymatically active uPA were determined
in malignant tissues of various origin [2, 3]. However, determination of
enzymatic activity of uPA in tissue extracts is not always an exact measure
of uPA content. Tumor tissues do also contain the inhibitors PAI-1 and
PAI-2. Disruption and solubilization of tumor cells during preparation of
tissue extracts free both uPA and its inhibitor (PAI-1/-2) which will then
lead to the formation of an enzymatically inactive but artificial uPA–PAI-1/-
2 complex. Most of the uPA in tissues is present in its enzymatically inactive

Table 2. Assessment of uPA-receptor (uPA-R) on cells and in solution

Reaction	Technique(s)	Reference(s)
Binding of radiolabeled uPA[a] to fluid-phase uPA-R or cell-associated uPA-R	Chemical crosslinking, equilibrium binding	21, 58, 61, 63, 118, 119
Reaction of receptor-bound uPA on single cells or tissue cells with: a) Fluorescent antibody to uPA b) Antibody to uPA and then fluoresceinated antibody to immunoglobulin	Fluorescence microscopy, laser scan microscopy, flow cytofluorometry	17, 29, 31, 47, 59
Binding of fluoresceinated uPA to cell-associated uPA-R	Laser scan microscopy, flow cytofluorometry	7, 29, 31, 117
Binding of fluorescent microspheres conjugated with an antibody directed to uPA-R	Light microscopy, fluorescence microscopy, laser scan microscopy, flow cytofluorometry, light scatter analysis	48
Binding of ferritin uPA to cell-associated uPA-R	Electron microscopy	19, 47, 62
Reaction of uPA-R in single cells or tissue cells with antibody to uPA-R and then with a second antibody to immunoglobulin (enzyme-labeled)	Immunocytochemistry, immunohistochemistry	Ronne and Dano, unpublished
Adsorption of uPA-R to polymer membrane, then reaction with antibody to uPA-R (conjugated with enzyme or biotin)	Immunodot	48
SDS-PAGE of uPA-R, then electrotransfer to polymer membrane and reaction with antibody to uPA-R (conjugated with enzyme or biotin)	Western blot	18, 63
Adsorption of antibody to uPA-R (catching antibody) to microtiter plate, then reaction with uPA. Detection by second antibody to uPA-R (conjugated with an enzyme or biotin)	Enzyme-linked immunosorbent assay	Ronne and Dano, unpublished

[a] pro-uPA, HMW-uPA, LMW-uPA, ATF, Kringle, GFD, synthetic uPA peptides.

Table 3a. Cancer of the breast, uterus and ovary: determination of uPA in plasma, ascitic fluid or cancer tissue

Type of cancer	Author	Year of publication	Reference
Breast	Burtin et al.	1987	65
	Clavel et al.	1986, 1988	66, 67
	Colombi et al.	1984	125
	Constantini et al.	1991	128
	Duffy et al.	1990, 1988, 1990, 1991	45, 68, 69, 71
	Evers et al.	1982	72
	Graeff et al.	1991	73
	Grondahl-Hansen et al.	1988	74
	Hayashi and Yamada	1985	75
	Jänicke et al.	1990, 1991, 1989	41, 42, 44
	Layer et al.	1987, 1987	76, 77
	McGregor et al.	1988	78
	Mira-y-Lopez et al.	1991	79
	Needham et al.	1987, 1988	80, 81
	O'Gradey et al.	1985	70
	Sappino et al.	1987	82
	Schmitt et al.	1991, 1990, 1990	10, 43, 46
	Sherman et al.	1980	83
	Tissot et al.	1984	84
Cervix uteri	Camiolo et al.	1986	116
	Koelbl et al.	1988	85
	Larsson et al.	1987	86
	Saito et al.	1990	87
Endometrium	Bulletti et al.	1991	60
	Koelbl et al.	1988	85
	Saito et al.	1990	87
	Soszka and Olszewski	1986	114
	Whitney et al.	1985	113
Ovary	Astedt and Holmberg	1976	40
	Camiolo et al.	1986	116
	Saito et al.	1990	87

pro-uPA form. Therefore quantification of uPA content in tumor tissues by ELISA has become an established laboratory method.

Compared with other tumor-associated antigens, the content of uPA in cancer tissues is rather low, in the range of a few nanograms per milligram of protein. Two high-affinity mAbs directed to different epitopes of uPA (#377 and #394, American Diagnostica) were selected by us to develop a highly sensitive uPA-ELISA (lower detection limit of 10 pg/ml). MAb #394 is also a useful tool to detect uPA antigen in formalin-fixed, paraffin-embedded breast cancer tissues by immunohistochemical techniques [41, 42]. MAb #394, which is directed to the B-chain of the uPA molecule, stained pro-uPA/uPA not only in the cytoplasm but also on the plasma membrane of breast cancer tumor cells, reflecting receptor-bound pro-uPA/

Table 3b. Cancer of the gastrointestinal tract: determination of uPA in plasma or cancer tissue

Type of cancer	Author	Year of publication	Reference
Colon	Burtin et al.	1985	88
	Camiolo et al.	1986	122
	Corasanti et al.	1980	120
	De Bruin et al.	1987, 1987, 1988, 1986	89, 90, 109, 124
	Fukao et al.	1991	64
	Gelister et al.	1986	123
	Grondahl-Hansen et al.	1991	50
	Harvey et al.	1988	92
	Huber et al.	1988	93
	Kirchheimer et al.	1987	94
	Kogha et al.	1985, 1989	52, 53
	Marian et al.	1990	95
	Markus et al.	1983	121
	Nishino et al.	1989	55
	Pyke et al.	1991	96
	Sier et al.	1991	51
	Sim et al.	1988	97
	Tanaka et al.	1991	98
	Tissot et al.	1984	84
	Verspaget et al.	1989	99
Gallbladder	Kirchheimer et al.	1987	94
	Huber et al.	1988	93
Esophagus	Nishino et al.	1989	55
Pancreas	Kirchheimer et al.	1987	94
Stomach	Fukao et al.	1991	64
	Huber et al.	1988	93
	Kirchheimer et al.	1987	94
	Nishino et al.		55
	Takai et al.	1991	100
	Tanaka et al.	1991	98

Table 3c. Cancer of the urogenital tract: determination of uPA in plasma or cancer tissue

Type of cancer	Author	Year of publication	Reference
Bladder	Gorelik et al.	1990	101
	Hasui et al.	1989	102
	Mitsubayashi et al.	1987	127
Kidney	Hata et al.	1989	111
	Kirchheimer et al.	1985	103
Prostate	Hienert et al.	1988	104, 129
	Kirchheimer et al.	1984, 1985	105, 106
	Wojtukiewicz et al.	1991	107

Table 3d. Cancer of the lung and the liver: determination of uPA in plasma or cancer tissue

Type of cancer	Author	Year of publication	Reference
Lung	Hayashi and Yamada	1990	126
	Markus et al.	1980	108
	Oka et al.	1991	115
	Sappino et al.	1987	82
	Veale et al.	1990	110
	Wajima	1991	56
Liver	Huber et al.	1988	93
	Kirchheimer et al.	1985	112
	Wojtukiewicz et al.	1990	91

Table 4. Quantitative assessment of uPA, tPA, PAI-1 and cathepsin D in breast cancer tissue extracts

Antigen	Mean	SD	Median
uPA[a]	3.04 (0.25)	2.52 (0.14)	2.38 (0.24) µg/mg protein
tPA[a]	17.87 (11.03)	21.04 (13.35)	9.39 (6.60) µg/mg protein
PAI-1[a]	1.75 (0.07)	2.72 (0.08)	1.03 (0.05) µg/mg protein
Cathepsin D	53.52 (11.09)	37.15 (17.34)	44.47 (5.99) p moles/mg protein

[a] in the presence of 1% (v/v) of the nonionic detergent Triton X-100 [41].

Breast cancer (primary tumors; $n = 274$) in italic; normal breast and benign diseases (fibroadenoma, mastopathy; $n = 27$) in parentheses.

uPA. The uPA-ELISA detected a statistically significant difference in pro-uPA/uPA content in breast cancer tissue extracts compared with normal breast tissue (Table 4) [10, 42, 43]. Breast cancer tissue extracts contained about 12 times more pro-uPA/uPA antigen than benign breast tissues. PAI-1 and cathepsin D antigen were also elevated (25 times more PAI-1 and 5 times more cathepsin D). No statistically significant difference between tPA antigen in breast cancer tissue and normal tissues was observed [10, 42, 43]. Pro-uPA/uPA antigen was also quantified in the plasma of breast cancer patients. Again, no statistically significant differences in pro-uPA/uPA levels were detected in plasma of patients with primary breast cancer, breast cancer patients who developed metastases after surgery of the primary tumor and a healthy control group [43]. Elevated levels of uPA antigen have also been detected in tumor extracts of other solid tumors (Table 3).

Clinical Relevance of uPA

Measurement of tumor-associated proteases has gained considerable importance during the past 3 years as it has become evident that, in addition to histomorphological features of the tumor and the established markers for cell proliferation, proteases may serve as independent predictors of tumor invasion and metastasis. Such factors are urgently needed to predict the outcome of the disease, especially in node-negative breast cancer patients. Cathepsin D, a lysosomal protease which is produced by many normal cells but secreted in excess by breast cancer cells, has been shown in breast cancer patients to be a valuable prognostic indicator to predict shorter overall survival and higher risk of relapse [33–35]. Likewise, breast cancer patients with relatively high pro-uPA/uPA content showed statistically significantly shorter relapse-free survival and shorter overall survival than those patients having tumors with lower pro-uPA/uPA levels [41, 42, 44, 45]. Determination of cathepsin D or pro-uPA/uPA content in breast cancer tissue extracts does not select the same group of patients, as cathepsin D and uPA are not correlated [10, 42].

The clinical course of breast cancer disease is associated not only with a high content of either cathepsin D or uPA but also with increased PAI-1 [42]. Patients with high PAI-1 antigen content in their primary tumor have a significantly higher relapse rate than patients with a low PAI-1 content, similar to uPA. We have shown recently that, including uPA and PAI-1 in a multivariate analysis, the two factors are of independent prognostic value for predicting the outcome of the disease [42]. Moreover, using a combination of both independent factors, uPA and PAI-1, the group of patients at high risk for relapse could be divided even further. Patients with tumors exhibiting high uPA and high PAI-1 content carry a maximal risk of relapse. Thus, by combination of uPA and PAI-1 a more individualized estimation of prognosis will become possible in breast cancer patients. In the future this may be of great importance for the clinical decision as to whether adjuvant chemotherapy should be given or not. In contrast to uPA, tPA is not significantly elevated in breast cancer primary tissues and therefore is not an indicator of poor prognosis [11, 42, 46]. Nevertheless, determination of tPA antigen may serve as an indicator of good prognosis in breast cancer, as tPA secretion by tumor cells is modulated by estrogen-mediated stimulation of tumor cells. tPA content in breast cancer tissue is thus related to functional steroid hormone receptors. For the proteases cathepsin B and L and collagenase IV, which are also elevated in breast cancer tissues, no detailed clinical studies concerning the course of the disease have yet been published.

Acknowledgments. This work was supported by the Wilhelm-Sander-Stiftung and the Deutsche Forschungsgemeinschaft (Sonderforschungsbereich 207, Projects F9 and G10). Dr. Moniwa is a visiting investigator supported by the Japanese government. We thank Dr. W.A. Günzler (Grünenthal, Stolberg,

FRG) for providing recombinant pro-uPA. The expert technical assistance of B. Münch, D. Hellmann and E. Sedlaczek was highly appreciated.

References

1. Bachmann F (1987) Thrombosis and haemostasis. In: Verstraete M, Lijnen HR, Arnout J (eds) International Society on Thrombosis and Haemostasis. Leuven University Press, Leuven, pp 227–265
2. Dano K, Andreasen PA, Grondahl-Hansen J, Kristensen P, Nielsen LS, Skriver L (1985) Plasminogen activators, tissue degradation, and cancer. Adv Cancer Res. 44:139–266
3. Markus G (1988) The relevance of plasminogen activators to neoplastic growth. Enzyme 40:158–172
4. Blasi F, Stoppelli MP, Cubellis V (1986) The receptor for urokinase-plasminogen activator. J Cell Biochem 32:179–186
5. Blasi F (1988) A surface receptor for urokinase plasminogen activator: a link between the cytoskeleton and the extracellular matrix. Protoplasma 145, 95–98
6. Blasi F (1988) Surface receptors for urokinase plasminogen activator. Fibrinolysis 2:73–84
7. Kobayashi H, Schmitt M, Goretzki L, Chucholowski N, Calvete J, Kramer M, Günzler WA, Jänicke F, Graeff H (1990) Cathepsin B efficiently activates the soluble and the tumor cell receptor-bound form of the proenzyme urokinase-type plasminogen activator (pro-uPA). J Biol Chem 266:5147–5152
8. Ichinose A, Fujikawa K, Suyama T (1986) The activation of pro-urokinase by plasma kallikrein and its inactivation by thrombin. J Biol Chem 261:3486–3489
9. Schmitt M, Kanayama N, Henschen A, Hollrieder A, Hafter R, Gulba D, Jänicke F, Graeff H (1989) Elastase released from human granulocytes stimulated with N-formyl-chemotactic peptide prevents activation of tumor cell prourokinase. FEBS Lett 255:83–88
10. Schmitt M, Goretzki L, Jänicke F, Calvete J, Eulitz M, Kobayashi H, Chucholowski N, Graeff H (1991) Biological and clinical relevance of the urokinase-type plasminogen activator (uPA) in breast cancer. Biomed Biochim Acta 50:737–741
11. Duffy MJ, O'Grady P, Devaney D, O'Siorain L, Fennelly JJ, Lijnen HR (1988) Tissue-type plasminogen activator, a new prognostic marker in breast cancer. Cancer Res 48:1348–1349
12. Roldan AL, Cubellis MV, Masucci MT, Behrendt W, Lund LR, Dano K, Appella E, Blasi F (1990) Cloning and expression of the receptor for human urokinase plasminogen activator, a central molecule in cell surface, plasmin dependent proteolysis. EMBO J 9:467–474
13. Kirchheimer JC, Binder BR (1991) Function of receptor-bound urokinase. Semin Thromb Hemost 17:246–250
14. Appella E, Robinson EA, Ullrich SJ, Stoppelli MP, Corti A, Casani G, Blasi F (1987) The receptor-binding sequence of urokinase. A biological function for the growth-factor module of proteases. J Biol Chem 262:4437–4440
15. Chucholowski N, Schmitt M, Goretzki L, Schüren E, Moniwa N, Weidle U, Kramer M, Wagner D, Jänicke F, Graeff H (1992) Flow cytofluorometry in tumor cell analysis. Biochem Soc Trans 20:208–216
16. Vassalli JD, Baccino D, Belin D (1985) A cellular binding site for the Mr 55 000 form of the human plasminogen activator, urokinase. J Cell Biol 100:86–92

17. Stoppelli MP, Tacchetti C, Cubellis MV, Corti A, Hearing VJ, Cassani G, Appella E, Blasi F (1986) Autocrine saturation of pro-urokinase receptors on human A431 cells. Cell 45:675–684

18. Chucholowski N, Schmitt M, Rettenberger P, Schüren E, Moniwa N, Goretzki L, Wilhelm O, Weidle U, Jänicke F and Graeff H (1992) Flow cytofluorometric analysis of the urokinase receptor (uPA-R) on tumor cells by fluorescent uPA-ligand or monoclonal antibody #3936 Fibrinolysis 6, Suppl. 4:95–102

19. Estreicher A, Muhlhauser J, Carpentier JL, Orci L, Vassalli JD (1990) The receptor for urokinase type plasminogen activator polarizes expression of the protease to the leading edge of migrating monocytes and promotes degradation of enzyme inhibitor complexes. J Cell Biol 111:783–792

20. Kirchheimer JC, Nong YH, Remold HG (1988) IFN-gamma, tumor necrosis factor-alpha, and urokinase regulate the expression of urokinase receptors on human monocytes. J Immunol 141:4229–4234

21. Stoppelli MP, Corti A, Soffientini A, Cassani G, Blasi F and Assoian RK (1985) Differentiation-enhanced binding of the amino-terminal fragment of human urokinase plasminogen activator to a specific receptor on U937 monocytes. Proc Natl Acad Sci USA 82:4939–4943

22. Nielsen LS, Kellermann GM, Behrendt N, Picone R, Dano K, Blasi F (1988) A 55 000–60 000 Mr receptor protein for urokinase-type plasminogen activator. Identification in human tumor cell lines and partial purification. J Biol Chem 263:2358–2363

23. Behrendt N, Ronne E, Ploug M, Petri T, Lober D, Nielsen LS, Schleuning WD, Blasi F, Appella E, Dano K (1990) The human receptor for urokinase plasminogen activator. NH2-terminal amino acid sequence and glycosylation variants. J Biol Chem 265:6453–6460

24. Ploug M, Ronne E, Behrendt N, Jensen AL, Blasi F, Dano K (1991) Cellular receptor for urokinase plasminogen activator: carboxyl-terminal processing and membrane anchoring by glycosyl-phosphatidylinositol. J Biol Chem 266:1926–1933

25. Ploug M, Behrendt N, Lober D, Dano K (1991) Protein structure and membrane anchorage of the cellular receptor for urokinase-type plasminogen activator. Semin Thromb Hemost 17:183–193

26. Vassalli J-D, Belin D (1987) Amiloride selectively inhibits the urokinase-type plasminogen activator. FEBS Lett 214:187–191

27. Schneider J (1990) Interactions of saruplase with acetylsalicylic acid, heparin, glyceryl trinitrate, tranexamic acid and aprotinin in a rabbit pulmonary thrombosis model. Arzneimittelforschung 40:1180–1184

28. Stephens RW, Pöllänen J, Tapiovaara H, Leung KC, Sim PS, Salonen EM, Ronne E, Behrendt N, Dano K, Vaheri A (1989) Activation of pro-urokinase and plasminogen on human sarcoma cells: a proteolytic system with surface-bound reactants. J Cell Biol 108:1985–1987

29. Schmitt M, Chucholowski N, Busch E, Hellmann D, Wagner B, Goretzki L, Jänicke F, Günzler WA, Graeff H (1991) Fluorescent probes to assess the receptor for the urokinase-type plasminogen activator on tumor cells. Sem Thromb Hemost 17:291–302

30. Miles LA, Plow EF (1988) Plasminogen receptors: ubiquitous sites for cellular regulation of fibrinolysis. Fibrinolysis 2:61–71

31. Lu H, Mirshahi M, Krief P, Soria C, Soria J, Mishal Z, Bertrand O, Perrot JY, Li H, Picot C (1988) Parallel induction of fibrinolysis and receptors for plasminogen and urokinase by interferon gamma on U937 cells. Biochem Biophys Res Commun 155:418–422

32. Appella E, Blasi F (1987) The growth factor module of urokinase is the binding sequence for its receptor. Ann NY Acad Sci 511:192–195

144   M. Schmitt et al.

33. Spyratos F, Maudelonde T, Brouillet J-P, Brunet M, Defrenne A, Andrieu C, Hacene K, Desplaces A, Rouesse J, Rochefort H (1989) Cathepsin D: an independent prognostic factor for metastasis of breast cancer. Lancet 8672:1115–1118
34. Tandon AT, Clark GM, Chamness GC, Chrigwin JM, McGuire WL (1990) Cathepsin D and prognosis in breast cancer. N Engl J Med 322:297–302
35. Thorpe SM, Rochefort H, Garcia M, Freiss G, Christensen IJ, Khalaf S, Paolucci F, Pau B, Rasmussen BB, Rose C (1989) Association between high concentrations of Mr 52000 cathepsin D and poor prognosis in primary human breast cancer. Cancer Res 49:6008–6014
36. Sloane BF, Dunn JR, Honn KV (1981) Lysosomal cathepsin B: correlation with metastatic potential. Science 212:1151–1153
37. Trygvasson K, (1989) Extracellular matrix and its enzymatic degradation in tumor invasion. In: Liotta LA (ed) Influence of tumor development on the host. Kluwer, Dordrecht, pp 72–83
38. Lah TT, Buck MR, Honn KV, Crissman JD, Rao NC, Liotta LA, Sloane BF (1989) Degradation of laminin by tumor cathepsin B. Clin Exp Metastasis 7:461–468
39. Sloane BF, Ryan RE, Rozhin J, Lah TT, Crissman JD, Honn KV (1986) Cathepsin B: a proteinase linked to metastases. In: Proteinases in inflammation and tumor invasion. de Gruyter, Berlin, pp 93–106
40. Astedt B, Holmberg L (1976) Immunological identity of urokinase and ovarian carcinoma plasminogen activator released in tissue culture. Nature 261:595–597
41. Jänicke F, Schmitt M, Hafter R, Hollrieder A, Babic R, Ulm K, Gössner W, Graeff H (1990) The urokinase-type plasminogen activator (u-PA) is a potent predictor of early relapse in breast cancer. Fibrinolysis 4:69–78
42. Jänicke F, Schmitt M, Graeff H (1991) Clinical relevance of the urokinase-type and tissue-type plasminogen activators and of their type 1 inhibitor in breast cancer. Semin Thromb Hemost 17:303–312
43. Schmitt M, Jänicke F, Hafter R, Hollrieder A, Kanayama N, Gulba D, Graeff H (1990) Tumor-associated fibrinolysis in human breast cancer: Detection and quantitation of the urokinase-type plasminogen activator (uPA) by ELISA and immunohistochemistry. In: Matsuda M, Iwanaga S, Takada A, Henschen A (eds) Fibrinogen:4. Current basic and clinical aspects. Elsevier, Amsterdam, pp 213–222
44. Jänicke F, Schmitt M, Ulm K, Gössner W, Graeff H (1989) Urokinase-type plasminogen activator antigen and early relapse in breast cancer. Lancet II 8670:1049
45. Duffy MJ, Reilley D, O'Sullivan C, O'Higgins N, Fennelly JJ, Andreasen P (1990) Urokinase-plasminogen activator, a new and independent prognostic marker in breast cancer. Cancer Res 50:6827–6829
46. Schmitt M, Jänicke F, Graeff H (1990) Tumor-associated fibrinolysis: The prognostic relevance of plasminogen activators uPA and tPA in human breast cancer. Blood Coagulation Fibrinolysis 1:695–702
47. Pöllänen J, Hedman K, Nielsen LS, Dano K, Vaheri A (1988) Ultrastructural localization of plasma membrane-associated urokinase-type plasminogen activator at focal contacts. J Cell Biol 106:87–95
48. Schmitt M, Jänicke F, Graeff H (1992) Tumor-associated proteases. Fibrinolysis 6, Suppl. 4:3–26
49. Grondahl-Hansen J, Kirkeby LT, Ralfkiaer E, Kristensen P, Lund LR, Dano K (1989) Urokinase-type plasminogen activator in endothelial cells curing acute inflammation of the appendix. Am J Pathol 135:631–636
50. Grondahl-Hansen J, Ralfkiaer E, Kirkeby LT, Kristensen P, Lund LR, Dano K (1991) Localization of urokinase-type plasminogen activator in stromal cells in adenocarcinomas of the colon in humans. Am J Pathol 138:111–117

51. Sier CFM, Fellbaum C, Verspaget HW, Schmitt M, Griffioen G, Graeff H, Höfler H, Lamers CB (1991) Immunolocalization of urokinase-type plasminogen activator (uPA) in adenomas and carcinomas of the human colon. Histopathology 19:231–237

52. Kohga S, Harvey SR, Weaver RM, Markus G (1985) Localization of plasminogen activators in human colon cancer by immunoperoxidase staining. Cancer Res 45:1787–1796

53. Kohga S, Harvey SR, Suzumiya J, Sumiyoshi A, Markus G (1989) Comparison of the immunohistochemical localisation of urokinase in normal and cancerous human colon tissue. Fibrinolysis 3:17–22

54. Kwaan C, Keer HN, Radosevich JA, Cajot JF, Ernst R (1991) Components of the plasminogen-plasmin system in human tumor cell lines. Semin Thromb Hemost 17:175–182

55. Nishino N, Aoki K, Tokura Y, Sakagushi S, Takada Y, Takada A (1989) The urokinase type plasminogen activator in cancer of digestive tracts. Thromb Res 50:527–535

56. Wajima T (1991) Fibrinolytic profiles in patients with small cell carcinoma of the lung. Semin Thromb Hemost 17:280–285

57. Brunner G, Simon MM, Kramer MD (1990) Activation of pro-urokinase by the human T cell-associated serine proteinase HuTSP-1. FEBS Lett 260:141–144

58. Schlechte W, Brittain M, Boyd D (1990) Invasion of extracellular matrix by cultured colon cells: dependence on urokinase receptor display. Cancer Commun 2:173–179

59. Niedballa MJ, Bajetta S, Carbone R, Satorelli AC (1990) Regulation of human squamous cell carcinoma plasma membrane associated urokinase plasminogen activator by epidermal growth factor. Cancer Commun 2:317–324

60. Bulletti C, Jasonni VM, Polli V, Cappuccini F, Galassi A, Flamigni C (1991) Basement membrane in human endometrium: possible role of proteolytic enzymes in developing hyperplasia and carcinoma. Ann NY Acad Sci 622:376–382

61. Masucci MT, Pedersen N, Blasi F (1991) A soluble, ligand binding mutant of the human urokinase plasminogen activator receptor. J Biol Chem 266:8655–8658

62. Dini G, Fibbi G, Pasquali F, del Rosso M (1985) Plasminogen activator: morphological evidence of binding, internalization and delivery to lysosomes in 3T3 mouse fibroblasts. Histochem J 17:333–341

63. Pedersen N, Ronne E, Schmitt M, Dano K, Hoyer-Hansen G, Kuhn W, Jänicke F, Blasi F (1991) Presence of a water-soluble unoccupied form of the urokinase-receptor in the ascitic fluid and plasma of patients with ovarian cancer. J Clin Invest (submitted)

64. Fukao H, Tanaka N, Ueshima S, Okada T, Yasutomi M, Matsuo O (1991) Plasminogen activator inhibitor in stomach and colorectal carcinoma. Semin Thromb Hemost 17:276–279

65. Burtin P, Chavanel G, André-Bougaran J, Gentile A (1987) The plasmin system in human adenocarcinomas and their metastases. A comparative immunofluorescence study. Int J Cancer 39:170–178

66. Clavel C, Chavanel G, Birembaut P (1986) Detection of the plasmin system in human mammary pathology using immunofluorescence. Cancer Res 46:5743–5747

67. Clavel C, Birembaut P (1988) Proteases et carcinomes mammaires. Ann Pathol 8:20–24

68. Duffy M, O'Grady P, Devaney D, O'Siorain L, Fennelly JJ, Lijnen HJ (1988) Urokinase-plasminogen activator, a marker for aggressive breast carcinomas. Cancer 62:531–533

69. Duffy MJ, Reilly D, O'Sullivan C, O'Higgins N, Fennelly JJ (1990) Urokinase plasminogen activator and prognosis in breast cancer. Lancet 335/8681:108
70. O'Grady P, Lijnen HR, Duffy MJ (1985) Multiple forms of plasminogen activator in breast tumors. Cancer Res 45:6216–6218
71. Duffy MJ, Brouillet JP, Reilley D, McDermott E, O'Higgins N, Fennelly JJ, Maudelonde TA, Rochefort H (1991) Cathepsin D concentration in breast cancer cytosols: correlation with biochemical, histological and clinical findings. Clin Chem 37:101–104
72. Evers JL, Patel J, Madeja JM, Schneider SL, Hobika GH, Camiolo SM, Markus GB (1982) Plasminogen activator activity and composition in human breast cancer. Cancer Res 42:219–226
73. Graeff H, Jänicke F, Schmitt M (1991) Klinische und prognostische Bedeutung tumorassoziierter Proteasen in der gynäkologischen Onkologie. Geburtshilfe Frauenheilkd 51:90–99
74. Grondahl-Hansen J, Agerlin M, Munkholm-Larsen P, Bach F, Nielsen LS, Dombernowsky P, Dano K (1988) Sensitive and specific enzyme-linked immunosorbent assay for urokinase-type plasminogen activator and its application to plasma from patients with breast cancer. J Lab Clin Invest 111:42–51
75. Hayashi S, Yamada K (1985) Urokinase-type plasminogen activator in ascites obtained from the patient with mammary cancer. Thromb Res 38:459–467
76. Layer GT, Cederholm-Williams SA, Gaffney PJ, Houlbrook S, Mahmoud M, Pattison M, Burnand K (1987) Urokinase – the enzyme responsible for invasion and metastasis in human breast carcinoma? Fibrinolysis 1:237–240
77. Layer GT, Burnand KG, Gaffney PJ, Cederholm-Williams SA, Mahmoud M, Houlbrook S, Pattison M (1987) Tissue plasminogen activators in breast cancer. Thromb Res 45:601–607
78. McGregor IR, Miller WR (1988) Urokinase and tissue-type plasminogen activators are present in breast cyst fluids. Eur J Cancer Clin Oncol 24:985–989
79. Mira-y-Lopez R, Osborne MP, DePalo AJ, Ossowski L (1991) Estradiol modulation of plasminogen activator production in organ cultures of human breast caracinomas: correlation with clinical outcome of anti-estrogen therapy. Int J Cancer 47:827–832
80. Needham GK, Sherbet GV, Farndon JR, Harris AL (1987) Binding of urokinase to specific receptor sites on human breast cancer membranes. Br J Cancer 55:13–16
81. Needham GK, Nicholson S, Angus B, Farndon JR, Harris AL (1988) Relation of membrane-bound tissue type and urokinase type plasminogen activators in human breast cancers to estrogen and epidermal growth factor receptors. Cancer Res 48:6603–6607
82. Sappino AS, Busson N, Belin D, Vassalli JD (1987) Increase of urokinase-type plasminogen activator gene expression in human lung and breast carcinomas. Cancer Res 45:601–607
83. Sherman MR, Tuazon FB, Miller LK (1980) Estrogen receptor cleavage and plasminogen activation by enzymes in human breast tumor cytosol. Endocrinology 106:1715–1727
84. Tissot J-D, Hauert J, Bachmann F (1984) Characterization of plasminogen activators from normal human breast and colon and from breast and colon carcinomas. Int J Cancer 34:295–302
85. Koelbl H, Kirchheimer JC, Tatra G, Christ G, Binder BR (1988) Increased plasma levels of urokinase-type plasminogen activator with endometrial and cervical cancer. Obstet Gynecol 72:252–256

86. Larsson G, Larsson A, Astedt B (1987) Tissue plasminogen activator and urokinase in normal, dysplastic and cancerous squamous epithelium of the uterine cervix. Thromb Haemost 58:822–826

87. Saito K, Nagashima M, Iwata M, Hamada H, Sumiyoshi K, Takada Y, Takada A (1990) The concentration of tissue type plasminogen activator and urokinase in plasma and tissues of patients with ovarian and uterine tumors. Thromb Res 58:355–366

88. Burtin P, Chavanel G, André J (1985) The plasmin system in human colonic tumors: an immunofluorescence study. Int J Cancer 35:307–314

89. De Bruin PA, Griffioen G, Verspaget HW, Verheijen JH, Lamers CB (1987) Plasminogen activators and tumor development in the human colon: activity levels in normal mucosa, adenomatous polyps and adenocarcinomas. Cancer Res 47:4654–4657

90. De Bruin PA, Griffioen G, Verspaget HW, Verheijen JH, Dooijeward G, van den Ingh HF, Lamers CB (1987) Plasminogen activator profiles in neoplastic tissues of the human colon. Cancer Res 4:4520–4524

91. Wojtukiewicz MZ, Zacharski LR, Memoli VA, Kisiel W, Kudryk BJ, Rousseau SM, Stump DC (1990) Abnormal regulation of coagulation/fibrinolysis in small cell carcinoma of the lung. Cancer 65:481–485

92. Harvey SR, Lawrence DD, Madeja JM, Abbey SJ, Markus G (1988) Secretion of plasminogen activators by human colorectal and gastric tumor explants. Clin Exp Metast 6:431–450

93. Huber K, Wojta J, Kirchheimer JC, Ermler D, Binder BR (1988) Plasminogen activators and plasminogen activator inhibitor in malignant ascitic fluid. Eur J Clin Invest 18:595–599

94. Kirchheimer JC, Huber K, Wagner O, Binder BR (1987) Pattern of fibrinolytic parameters in patients with gastrointestinal carcinomas. Br J Haematol 66:85–89

95. Marian B, Harvey S, Infante D, Markus G, Winawer S, Friedman E (1990) Urokinase secretion from human colon carcinomas induced by endogenous diglycerides. Cancer Res 50:2245–2250

96. Pyke C, Kristensen P, Ralfkiaer E, Grondahl-Hansen J, Eriksen J, Blasi F, Dano K (1991) Urokinase-type plasminogen activator is expressed in stromal cells and its receptor in cancer cells at invasive foci in human colon adenocarcinomas. Am J Pathol 138:1059–1067

97. Sim PS, Stephens RW, Fayle DR, Doe WF (1988) Urokinase-type plasminogen activator in colorectal carcinomas and adenomatous polyps: quantitative expression of active and proenzyme. Int J Cancer 42:483–488

98. Tanaka N, Fukao H, Ueshima S, Okada K, Yasutomi M, Matsuo O (1991) Plasminogen activator inhibitor 1 in human carcinoma tissues. Int J Cancer 48:481–484

99. Verspaget HW, Verheijen JH, de Bruin PA, Griffioen G, Lamers CB (1989) Plasminogen activators in (pre) malignant conditions of the colorectum. Eur J Cancer Clin Oncol 25:565–569

100. Takai S, Yamamura M, Tanaka K, Kawanishi H, Tsuji M, Nakane Y, Hioki K, Yamamoto M (1991) Plasminogen activators in human gastric cancers: correlation with DNA ploidy and immunohistochemical staining. Int J Cancer 48:20–27

101. Gorelik U, Lindner A, Mayer M (1990) Protease and plasminogen activity in human bladder carcinoma. Br J Urol 66:170–174

102. Hasui Y, Suzumiya J, Marutsuka K, Sumiyoshi A, Hashida S, Ishikawa E (1989) Comparative study of plasminogen activators in cancers and normal mucosae of human urinary bladder. Cancer Res 49:1067–1070

103. Kirchheimer JC, Pflüger H, Hienert G, Binder BR (1985) Increased urokinase activity to antigen ratio in human renal-cell carcinoma. Int J Cancer 35:737–741
104. Hienert G, Kirchheimer JC, Pflüger H, Binder BR (1988) Urokinase-type plasminogen activator as a marker for the formation of distant metastases in prostatic carcinomas. J Urol 140:1466–1469
105. Kirchheimer JC, Köller A, Binder BR (1984) Isolation and characterization of plasminogen activators from hyperplastic and malignant prostate tissue. Biochim Biophys Acta 797:256–265
106. Kirchheimer JC, Pflüger H, Ritschl P, Hienert G, Binder BR (1985) Plasminogen activator activity in bone metastases of prostatic carcinomas as compared to primary tumors. Invasion Metastasis 5:344–355
107. Wojtukiewicz MZ, Zacharski LR, Memoli VA, Kisiel W, Kudryk BJ, Moritz TE, Rousseau SM, Stump DC (1991) Fibrin formation on vessel walls in hyperplastic and malignant prostate tissue. Cancer 67:1377–1383
108. Markus G, Takita H, Camiolo SM, Corasanti JG, Evers JL, Hobika GH (1980) Content and characterization of plasminogen activators in human lung cancer and normal lung tissue. Cancer Res 40:841–848
109. De Bruin PAF, Griffioen G, Verspaget HW, Verheijen JH, Dooijewaard G, van den Ingh HF, Lamers CBHW (1988) Plasminogen activator profiles in neoplastic tissues of the human colon. Cancer Res 48:4520–4524
110. Veale D, Needham G, Harris AL (1990) Urokinase receptors in lung cancer and normal lung. Anticancer Res 10:417–421
111. Hata M (1989) The study of plasminogen activator in renal carcinoma with special remarks on urokinase type plasminogen activator. Nippon Hinyokika Gakkai Zasshi 80:1558–1565
112. Kirchheimer JC, Huber K, Polterauer P, Binder BR (1985) Urokinase antigen in plasma of patients with liver cirrhosis and hepatoma. Thromb Haemost 54:617–618
113. Whitney CW, Satyaswaroop PG, Mortel R (1985) Plasminogen activator activity in human endometrial carcinoma. Gynecol Oncol 22:97–104
114. Soszka T, Olszewski K (1986) Plasminogen activators and their inhibitors in normal, hyperplastic and carcinomatous human endometrium. Thromb Res 42:835–846
115. Oka T, Ishida T, Nishino T, Sugimachi K (1991) Immunohistochemical evidence of urokinase-type plasminogen activator in primary and metastatic tumors of pulmonary adenocarcinoma. Cancer Res 51:3522–3525
116. Camiolo SM, Markus G, Piver MS (1987) Plasminogen activator content of gynecological tumors and their metastases. Gynecol Oncol 26:364–373
117. Takahasi K, Gojobori T, Tanifuji M (1991) Two-color cytofluorometry and cellular properties of the urokinase receptor associated with a human metastatic carcinomatous cell line. Exp Cell Res 192:405–413
118. Plow EF, Freaney DE, Plescia J, Miles LA (1986) The plasminogen system and cell surfaces : evidence for plasminogen and urokinase receptors on the same cell type. J Cell Biol 103:2411–2420
119. Estreicher A, Wohlwend A, Belin D, Schleuning W-D, Vassalli J-D (1989) Characterization of the cellular binding site for the urokinase-type plasminogen activator. J Biol Chem 264:1180–1189
120. Corasanti JG, Celik C, Camiolo SM, Mittelman A, Evers JL, Barbasch A, Hobika GH, Markus G (1980) Plasminogen activator content of human colon tumors and normal mucosae: separation of enzymes and partial purification. J Natl Cancer Inst 65:345–35
121. Markus G, Camiolo SM, Kohga S, Madeja JM, Mittelman A (1983) Plasminogen activator secretion of human tumors in short-term organ culture, including a comparison of primary and metastatic colon tumors. Cancer Res 43:5517–5522

122. Camiolo SM, Greco WR (1986) Plasminogen activator content of human tumor and adjacent normal tissue measured with fibrin and non-fibrin assays. Cancer Res 46:1788–1794
123. Gelister JSK, Mamoud M, Lewin MR, Gaffney PJ, Boulos PB (1986) Plasminogen activators in human colorectal neoplasia. Br Med J 293:728–731
124. DeBruin PAF, Verspaget HW, Griffioen G, Verheijen JH, Lamers CBHW (1986) Plasminogen activator activity in colonic polyp carcinoma sequence. Gut 27:A1267
125. Colombi M, Barlati S, Magdalenai H, Fiszer-Szafarz B (1984) Relationship between multiple forms of plasminogen activator in human breast tumors and plasma and the presence of metastases in lymph nodes. Cancer Res 44: 2971–2975
126. Hayashi S, Yamada K (1990) Properties of urokinase type-plasminogen activator found in chest fluid. Nippon Ketsueki Gakkai Zasshi 53:105–115
127. Mitsubayashi S, Akiyama T, Kurita T, Okada K, Bando H, Sakai T, Matsuo O (1987) Plasminogen activators in bladder tumors. Urol Res 15:335–339
128. Constantini V, Zacharski LR, Memoli VA, Kudryk BJ, Rousseau SM, Stump DC (1991) Occurrence of components of fibrinolysis pathways in situ in neoplastic human breast tissue. Cancer Res 51:354–358
129. Hienert G, Kirchheimer JC, Christ G, Pflüger H, Binder BR (1988) Plasma urokinase-type plasminogen activator correlates to bone scintigraphy in prostatic carcinoma. Eur Urol 15:256–258

Analysis of Growth Regulatory Pathways in Human Neuro-Oncology

M. Westphal, W. Hamel, L. Anker, and H.-D. Herrmann

Abteilung für Neurochirurgie, Labor für Hirntumorbiologie,
Universitätskrankenhaus Eppendorf, Martinistraße 52, W-2000 Hamburg 20, FRG

Introduction

Considering the therapies which are available today for human gliomas, the most promising new developments are to be expected from advances in the understanding of glial cell biology together with the technical capacity to manipulate cells by biochemical or molecular biological means. Herein, the exploration of growth factor pathways that may be deregulated has provided new understanding of the disease as well as therapeutic perspectives. The concomitant expression of growth factor receptors and secretion of growth factors resulting in autocrine self-stimulation of cells, which is frequently found in oncology [1, 14], is also present in glial tumors and will be presented from the neuro-oncological aspect.

The Role of Growth Factors in Glial Oncology

Early clinical observations implied a role for growth factors in the biology of gliomas. Comparing histology and in vitro behavior of primary cultures of anaplastic gliomas and glioblastomas, it was frequently seen that histology does not allow the prediction of culture development and dynamics. Two glioblastomas which show all the necessary histological hallmarks, such as necrosis and endothelial cell proliferation, but behave differently in vitro are compared in Fig. 1. Some glioma cultures spontaneously differentiate in vitro and others continue to proliferate and finally become established as immortal cell lines. Such differences of behavior in vitro must be related to the capacity to sustain in vitro a milieu similar to the in vivo situation – probably by secretion of a complex combination of growth factors.

The hypothesis of growth factor secretion by glial tumors was first tested by adding tumor cyst fluids to primary cultures of gliomas or other indicator cell lines. This resulted in a stimulation of cell proliferation in all cases,

Wagener/Neumann (Eds.)
Molecular diagnostics of Cancer
© Springer-Verlag Berlin Heidelberg 1993

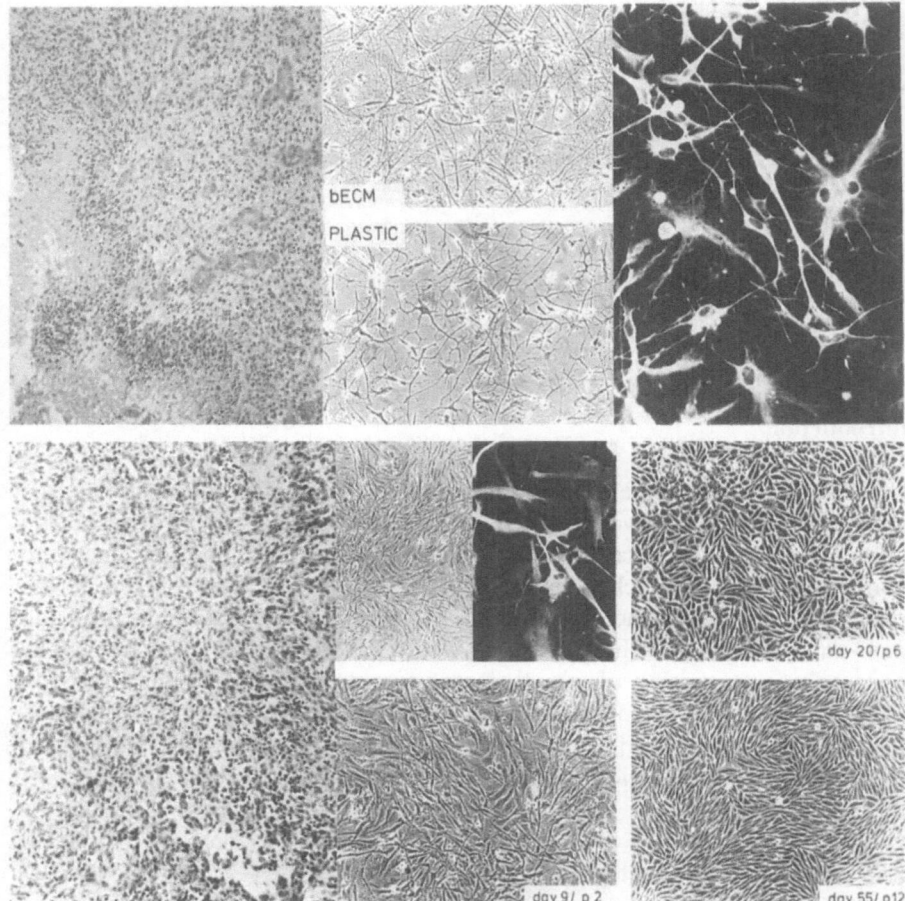

Fig. 1. Two glioblastomas and the corresponding cell cultures. *Upper panel*: Conventional histology and phase-contrast micrographs of cultured cells and immunostainings for glial fibrillary acidic protein (GFAP). This case differentiated in culture in both culture conditions and remained unchanged for 2 months. *Bottom panel*: Conventional histology and phase-contrast micrographs of the evolution in cell culture together with a GFAP stain in passage 1

proving that there is active growth factor secretion by tumors in vivo [56]. Next the autocrine hypothesis was tested in more stable systems such as cell lines. Twenty-four glioma-derived cell lines have been generated in our laboratory (Table 1). Twenty of these, including sublines or lines obtained after passage through the nude mouse, were tested for autocrine activation by adding conditioned serum-free medium to autologous cultures which had just been changed to serum-free medium. Dose-dependent stimulation of proliferation was seen in 50% of cell lines, while the others proliferated independently, as seen in two representative examples in Fig. 2 [59]. Of

Table 1. Growth factors and oncogenes in gliomas

Growth factor (receptor)[a]	Production by glioma	Corresponding oncogene
EGF/TGfa	Yes	
EGF-R	Often overexpressed	c-*erb*-B1
PDGF-A	Yes	
PDGF-B	Often overexpressed	c-*sis*
PDGF-Ra	Yes	
PDGF-Rb	Yes	
Acidic FGF	Yes	
Basic FGF	Yes	*hst, int*-2
FGF-R	Yes	
NGF	Yes	
NGF-R	??	*trk*

[a] A hyphenated R designates a receptor.

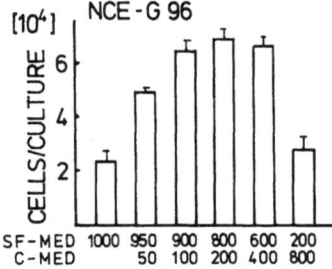 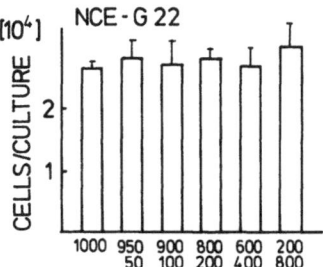

Fig. 2. Two cell lines exemplifying autocrine stimulation (NCE-G96) and independence of proliferation from autologous conditioned medium (*C-MED*; NCE-G22). The cells were seeded at 20 000 cells per well and then autologous conditioned serum-free medium (*Sf-MED*) was added in the indicated proportions. The cells were then left to grow and counted in a Coulter counter after 3 days

course, such results have to be interpreted with caution because they were obtained in cell lines adapted to the in vitro situation, and the actual proportion of cell cultures which are dependent on autocrine growth factor secretion may be higher before adaptation. The high proportion, however, of cell lines growing "independent" of autocrine growth factor secretion indicates that therapeutic strategies aimed at neutralizing this autoregulatory pathway may be doomed to failure in a significant proportion of patients. This was supported by our findings that the blockade of growth factor pathways by suramin, a polyanionic agent (200 µg/ml) which prevents epidermal growth factor (EGF), fibroblast growth factor (FGF) and platelet-derived growth factor (PDGF) from interacting with the cellular signalling mechanisms [16, 20, 30, 44], did so only incompletely in several cell lines [58].

Most of the known growth factor systems have been related to glial differentiation or the biology of gliomas [57] (Table 2). The systems analyzed in greatest depth are those of PDGF, FGF and EGF. In addition, insulin-like growth factors (IGFs) and, recently, nerve growth factor (NGF) have been linked with glial biology. Whereas there are true indications of a physiological and pathophysiological role of IGFs in glial development or neoplasia, the indications for the involvement of NGF are vague. NGF is produced by glioma cell lines (Hamel et al., unpublished data) but as yet there is no evidence that NGF receptors are present or transmit a growth stimulatory signal to glial cells. NGF has recently been shown to bind to the *trk* oncogene product which is related to the tyrosine kinase family of receptors but there is no evidence of *trk* expression in gliomas yet.

Table 2. Clinical data of glioma-derived cell lines

Cell line (NCE-G)	Patient's sex	Patient's age (years)	Histology	WHO grade	Location
22	F	56	Anaplastic astrocytoma	3	Medio-occipital
28	M	66	Gliosarcoma	4	Left parietal
44	F	64	Glioblastoma	4	Left parietal
55	M	65	Astrocytoma	3	Right occipitoparietal
59	M	66	Anaplastic astrocytoma	4	Left temporomedial
60	F	68	Anaplastic oligodendroglioma	4	Right parietotemporal
61	M	58	Anaplastic astrocytoma	4	Right temporo-occipital
62	F	4	Glioblastoma	4	Left temporoparietal
63	M	52	Glioblastoma	4	Right occipitoparietal
84	F	47	Anaplastic fibrillary	3	Left temporoparietal
96	F	9	Anaplastic astrocytoma	4	Left frontal
111	F	47	Glioblastoma	4	Left temporoparietal
112	M	51	Glioblastoma	4	Right temporal
118	F	54	Glioblastoma	4	Right frontal
120	M	59	Glioblastoma	4	Left temporoparietal
121	F	53	Glioblastoma, recurrent	4	Left occipitoparietal
122	M	50	Glioblastoma	4	Right frontal
123	M	68	Glioblastoma	4	Left parieto-occipital
124	M	58	Anaplastic astrocytoma	4	Right temporal
130	M	51	Anaplastic	4	Right temporal
140	M	59	Glioblastoma	4	Right occipital
141	M	52	Glioblastoma	4	Left frontal
142	M	58	Glioblastoma	4	Left temporal
167	M	50	Anaplastic astrocytoma	4	Right parietal

PDGFs are growth factors which have been shown to promote division of glial precursor cells during development and are responsible for determination of differentiation pathways or lineage selection of glial cells [40, 41]. PDGFs are made by gliomas, as are their receptors (PDGF-R). All possible forms of dimers between the A and B chains have been demonstrated in cell lines, and likewise expression of both genes has been seen [28]. PDGFs and PDGF-R constitute a strong autocrine loop but there are differences between the AA homodimer, on the one hand and the AB heterodimer and the BB-form, on the other, with the AA homodimer being a weaker mitogen [27]. Recent findings demonstrated correspondingly different forms of PDGF-R, an alpha and a beta form. Using molecular probes and in situ hybridization it appears, in original tumor tissues, as if the glial cells preferentially make PDGF-AA whereas the hyperplastic endothelial cells make PDGF-BB. The PDGF-AA produced by glioma cells interacts only with the A-type receptor, also located on the glioma cells. The PDGF-B type receptor, however, binds all forms of PDGF, is found on the endothelial cells and is suspected to be crucial for neovascularization. Thus, there may be two separate loops for the PDGF system within one tumor, as supected from in situ hybridization results (M. Nister, Asilomar conference communication). If this holds true, the in situ results are in contrast to in vitro results from cell cultures, in which binding of all forms of PDGFs could be found to cultured glioma cells. In this context we looked at gene expression of the two different PDGF receptors in our cell lines using northern blots and found that they are not made in parallel and also with great differences in quantity (Fig. 3).

The basic and acidic FGFs are of well-established relevance in glioma biology [10, 26, 48]. The factors have been shown to be produced, although there is no clear evidence for secretion as there is no appropriate signal peptide. However, it is speculated that secretion is accomplished by complexing of the FGFs to components of the extracellular matrix. From such a reservoir it might be released by proteolysis, but one must accepted that there is no primary role of FGFs as secreted, autocrine growth factors. Firm evidence has been produced for a role of at least acidic FGF in neovascularization. In developmetal biology, an immortalizing action of basic FGF on glial precursor cells together with PDGF has been observed [3].

The other growth factor system which has been equally extensively described is the EGF/EGF-R complex. Secretion of transforming growth factor alpha (TGFα), a homologue to EGF, is frequently seen in neoplasia and has also been firmly established for gliomas [28, 42]. Secretion of EGF is negligible. This became an interesting finding together with the fact that the corresponding receptor for both peptides, EGF-R, is often overexpressed in gliomas, a finding recorded in original tumor specimens [22, 49]. This overexpression was found to be due to gene amplification or gene rearrangements [2, 22]. In the meantime it has been established that EGF-R overexpression is a constant finding in about 50% of high-grade gliomas. The

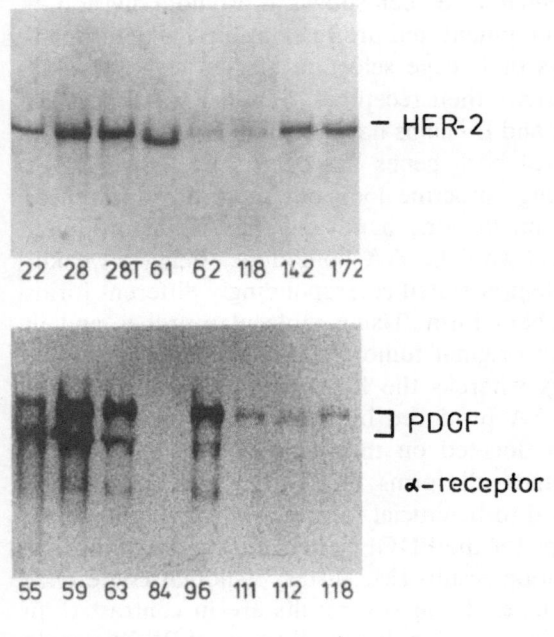

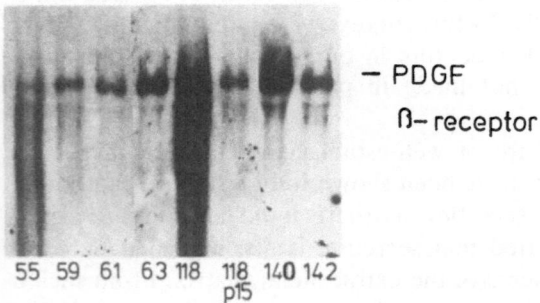

Fig. 3. Platelet-derived growth factor receptor gene expression analyzed by northern blot. Using probes for the PDGFα receptor, PDGFβ receptor and c-*erb*-B2 (HER-2), results with different cell lines (Table 1) are shown, indicating that the levels of expression differ among the cell lines. Only one cell line shown (underlined) is negative for gene expression of the PDGFα receptor. Those not marked turned out to have a weak signal after longer exposure (not shown). Equal amounts of RNA based on actin probing were loaded

gene for EGF-R is located on chromosome 7 and numerical increase has become an accepted cytogenetic feature of glioblastoma [15] (Fig. 4). For reasons of completeness it needs to be mentioned that the gene for the PDGF A chain is also located on chromosome 7 [45].

EGF has a broad spectrum of effects on glial cells as well as glioma cells. It regulates differentiation, proliferation and migration [7, 23, 43, 53]. From our own experiments and the reports of others we have to conclude that

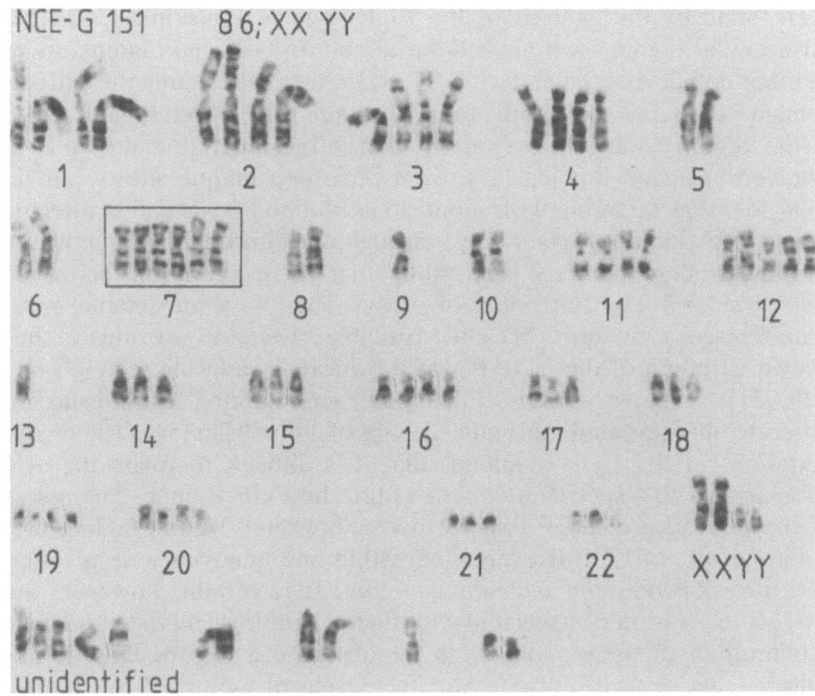

Fig. 4. Karyotype of human glioblastoma cell obtained by G-banding technique after 2 weeks in culture. The typical numeric increase of chromosome 7 (*box*) as well as characteristic losses of chromosomes 9, 10 and 13 is seen, together with several other random aberrations which are not derivatized in detail

EGF is not as strong a mitogen for glioma cells in culture as are the FGFs or PDGFs [55]. Nevertheless, overexpression of EGF-R is a strikingly constant feature and must serve a function even if it is not yet clear what that function is.

One of the first questions arising in this context is the possible molecular mechanism for gene activation and/or overexpression. Chromosomal translocations have been shown to cause oncogene activation [13], and genomic rearrangement and gene amplification have been shown for the EGF-R gene in a high proportion of gliomas in early studies. In later studies this could not be confirmed to occur as frequently as suspected, and it is usually not seen in cell lines. In our own series we looked at genomic rearrangement with Southern blots but found no restriction fragment size heterogeneities (Fig. 5B), indicating that the gene is most likely not expressed due to relocation to another transcriptional control unit within the genome. The analysis of the EGF-R protein, however, has revealed many anomalies, some of which may well be related to prolonged half-life or decreased rate of internalization and therefore less degradation. These mutants, which have been found by the analysis of the EGF-R gene transcripts, can affect the extracellular domain with or without alteration of ligand binding, as well as all other domains of the molecule [17, 47]. Mutants affecting the extracellular domain, e.g., the frequently found mutant with a deletion of 801 bases in the region coding for a part of the extracellular domain which is not involved in ligand binding [47], even provide a unique site of antigenicity ideal for drug targeting with monoclonal antibodies. Mutants affecting the intracellular kinase domain may, through constitutive kinase activation, be of transforming character [11]. Altogether these alterations show remarkable structural and functional diversity. They were detectable only after gene-transcript analysis became available, because previously the only known variation of the EGF-R was a truncated molecule secreted by A431 cells [51]. Our own studies with affinity cross-linking using disuccinimidyl suberate only revealed molecular species of 170 000 Da [54]. The diversity of mutations of the EGF-R implies that it is difficult to relate the origin of glioblastoma to a specific mutation within the EGF-R gene. The assessment of the relevance of EGF-R is further complicated by the finding of clonal heterogeneity of EGF-R expression within one tumor as well as differences of expression between individual tumors. It is certain, however, that the EGF-R expression is associated with higher histological grade and persistent proliferation in tissue culture, as we observed a loss of EGF-R in those cultures that ceased to proliferate after weeks or months [54].

Another oncogene-related molecule has been found to be of close similarity to the EGF-R. The product of the c-erb-B2 (HER-2) gene was also assumed to be an EGF-R binding protein, but mean while a specific ligand for the molecule has been isolated and completely described (H.M. Shepard, personal communication). It appears to be of relevance in other tumors such as mammary carcinoma and ovarian cancer but has only recently been analyzed in gliomas. When FACS data for EGF-R and HER-2 on cells from glioma cell lines are compared it appears that there is an inverse correlation, as has been shown for the two molecules in the other cancers. Overall, HER-2 expression was much rarer in glioma cell lines than the expression of EGF, but it was clearly demonstrable in some cases (Fig. 5). It was also of interest to see that all cell lines expressed HER-2 on the mRNA level but that not all cells were positive at the protein level, hinting at possible intracellular control mechanisms though sequestration or degradation.

With the fact becoming accepted that the EGF/TGFα and EGF-R system may not be as crucial for proliferation in gliomas as other growth factor systems, interest has shifted towards the analysis of EGF/EGF-R in migration or tumor invasion. Especially in the context of malignant gliomas this is very relevant, because any therapy comprising a local component such as topical application of immunotoxins, cytotoxins or cytokines is hampered by the fact that glioma cells migrate readily through the brain and can be found many centimeters away from the original site of the lesion [4]. It would thus

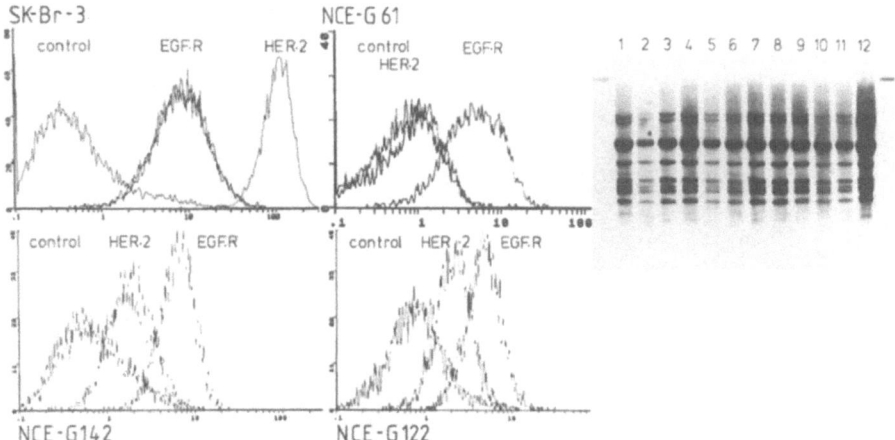

Fig. 5. *Left*: Detection of EGF-R and the c-*erb*-B2 (HER-2) expression on the cell surface by staining with monoclonal antibodies and subsequent analysis by fluorescence-activated cell separation (FACS). With cell line SK-Br-3 as reference, three different glioma cell lines (Table 1) were analyzed and showed relatively high expression of both receptors (NCE-G122 and NCE-G142) or only of EGF-R (NCE-G61). *Right*: Southern blot of genomic DNA isolated from 12 glioma cell lines, digested with *Eco*RI and probed with a cDNA recognizing c-*erb*-B1

be attractive to find mediators for the migratory activity, and EGF has already shown in early experiments that it might be of relevance in this respect to glia cells [53]. To study this aspect, multicellular spheroids are useful tools, and standardized migration assays have been developed [23]. Advantage is taken of the spontaneous formation of spheroids by cultured cells after they are placed on top of soft agar. These spheroids can be placed into culture dishes where they attach and can be treated with individual growth factors. In such an assay it could again be demonstrated that the response of individual cell lines to growth factors is as heterogeneous as it is in respect to proliferation (Fig. 6). However, it has become apparent that EGF is much more constantly efficacious in stimulating migration than in stimulating proliferation, and in the former activity it may match the importance of PDGF and the FGFs.

There is no doubt that growth factors are relevant to the pathophysiology of gliomas. They serve an autostimulatory purpose and mediate migration signals as well. However, from experiments with antagonization of individual growth factors as well as whole growth factor systems it appears as if the growth factors are not the transforming agents themselves. The addition of suramin for example, a substance which blocks the action of EGF, FGFs and PDGFs, is not sufficient to reverse the transformed phenotype or even inhibit proliferation [58]. Rather, it appears that the growth factors serve a role which is appropriate to the cellular program that is enacted at the time when a transforming event takes place in a cell. There is a great diversity

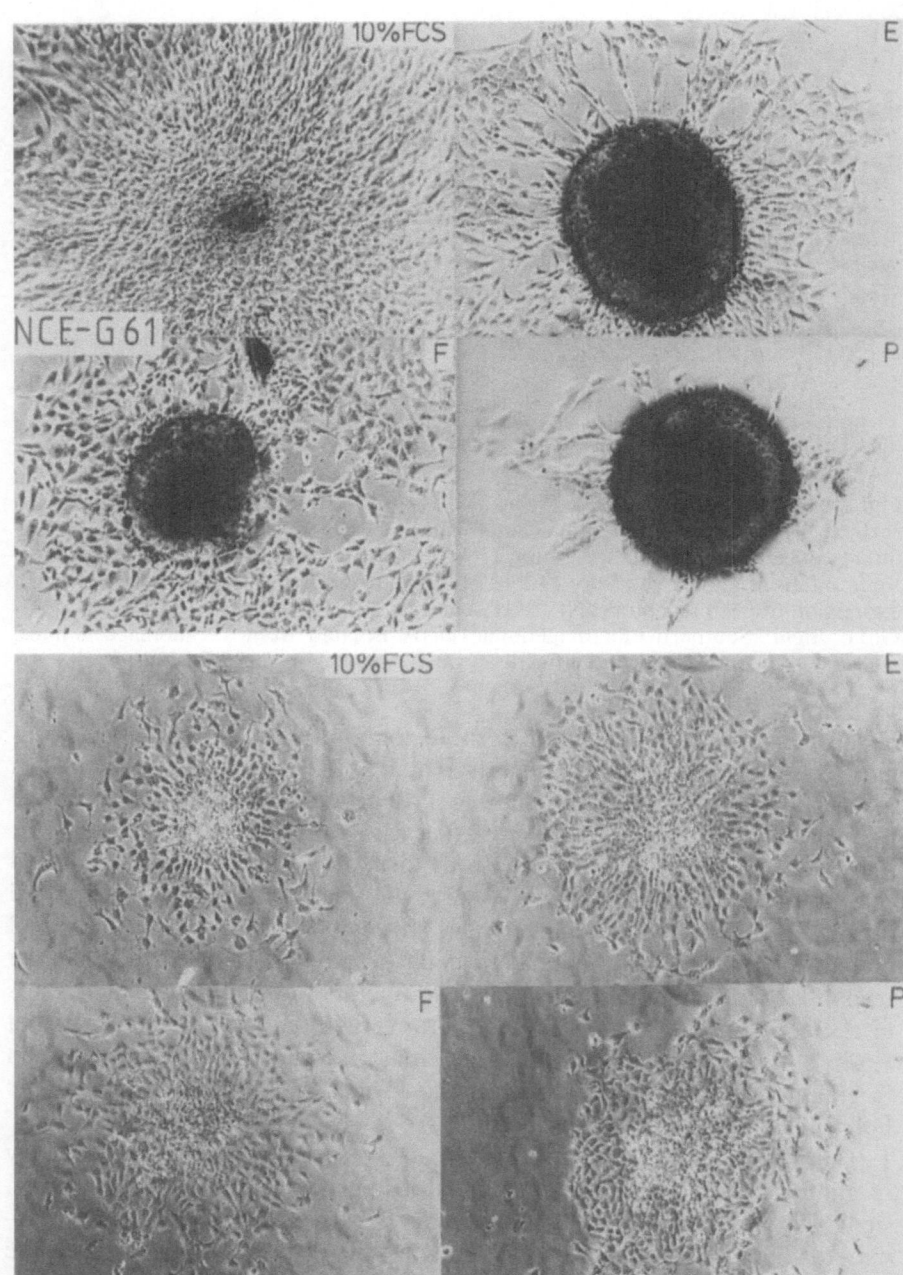

Fig. 6. The effect of growth factors (E, EGF 10^{-6} M; F, FGF 80 pg/ml; P, PDGF-AA 4 nM) on cell migration was tested in multicellular spheroids which were maintained in serum-free culture and assessed after 4 days of culture. *Top*, Spheroids from cell line NCE-G61; *bottom*, spheroids from cell line NCE-G96

of transforming processes, as had to be expected on the basis of clinical experience.

The Genetics of Glioma and the Role of Suppressor Genes

From daily clinical experience it is evident that glioblastoma, the most malignant form of glioma, arises either de novo with no sign of preexisting disease or will slowly develop from a low-grade lesion within a matter of years (Fig. 7). Thus two genetic mechanisms may be postulated: deregulation of a very fundamental cellular regulating gene and sequential loss of control mechanisms together with the acquisition of advantageous genes. Analyzing the genetic aberrations which have been shown to be involved in glioma development and progression, both components appear to be equally important: amplification and overexpression of advantageous genes and loss of control or "suppressor" genes [1, 52].

The only oncogene of which overexpression has been clearly shown to be associated with the development of brain tumors is the v-*sis* gene that encodes a PDGF-B like molecule and is homologous to the cellular c-*sis* [6]. FGF had to be modified to carry a signal peptide for secretion but then was able to tranform cells in culture. The EGF-R gene itself, when transfected as wild type into cultured cells, causes an EGF/TGFα-dependent transformation, in contrast to some of the deletion mutants which are constitutively activated. Overexpression of the EGF-R wild-type without constant stimulation by exogenous EGF appears to be without transforming effect [11, 46]. Thus, the focus of attention is shifting towards the suppressor genes, which have been shown to give clues in many types of neoplasia.

Loss of genetic information or mutations in control genes have been shown to be related to many tumors. Retinoblastoma is a classic case in which the loss of a gene on chromosome 13 is related to the familial form of the disease [8, 21]. The corresponding gene product, named pRB, is thought to take part in the regulation of gene transcription through interaction with transcription factors and probably serves as a fundamental nuclear regulator [5]. Another gene, located on chromosome 17p, has been shown to be frequently mutated or lost in cancer and is called the p53 gene as it encodes a protein of a molecular mass of 53 kDa, p53.

As far as gliomas are concerned, loss of heterozygosity for loci on chromosome 17p, indicating allelic loss, have been demonstrated [9, 18], as have frequent losses on chromosome 9 possibly involving the interferon genes [19]. The 17p alterations can be found very early in the glioma disease and can be detected in low-grade gliomas. Deletions of the interferon genes, however, are closely associated with the more anaplastic varieties. During progression to the rapidly recurring, highly malignant variants a very specific allelic loss of chromosome 10, which may or may not be associated with amplification of the EGF-R gene, is found [25]. Loss of chromosome 10 is

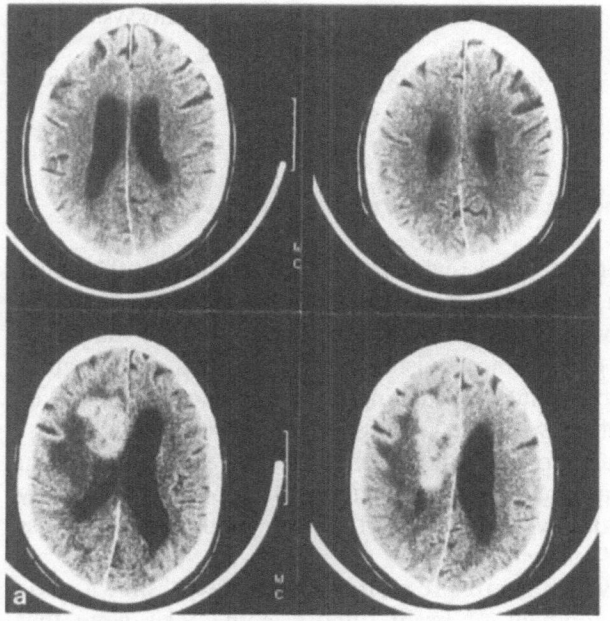

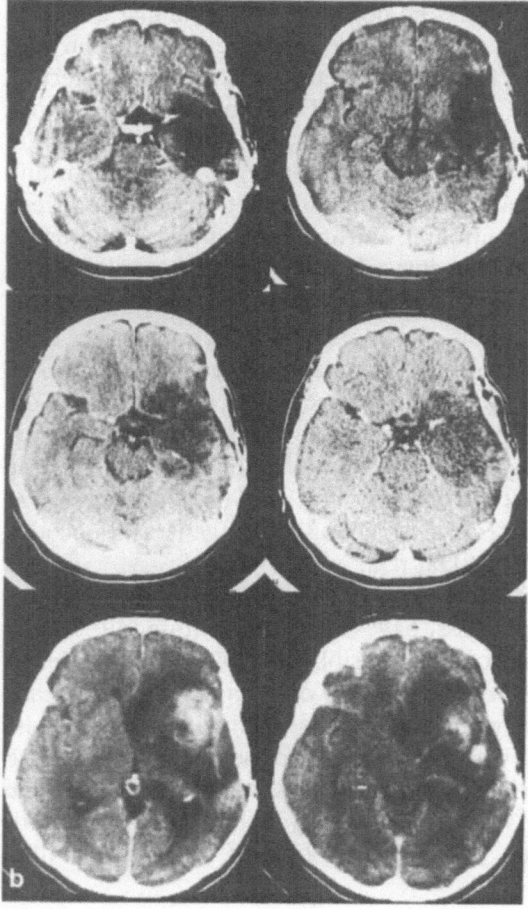

Fig. 7a,b. Series of computed tomography studies from two patients illustrated the different clinical courses of glioma. **a** A 50-year-old man had an investigation (*top*) because of recurrent episodes of sudden deafness; the findings were normal. Scans taken 6 months later because of dizziness and drowsiness (*bottom*) showed a tumor which turned out to be glioblastoma. **b** The slow progression of a glioma in a young woman, beginning as a low-grade lesion in 1985 (*top*) and 1987 (*middle*) and progressing to anaplastic glioma by 1991 (*bottom*)

also present in the primary glioblastomas which occur in the absence of symptoms for preexisting low-grade disease.

Mutations of p53 leading to loss of function of this gene are related to the loss of genetic information on chromosome 17p and are thus found frequently in gliomas [29]. There appears to be a whole range of possible point mutations, and most of these lead to an increased half-life of the molecule, which may be useful for detection as discussed in the legend to figure 8 (Fig. 8). Such analysis, however, has to be complemented by further gene analysis because a positive immunostaining may be deceptive (Fig. 8). We have so far analyzed only our cell lines, but these show mutations in 30% of cases and complete loss of detectable protein also in 30%, a higher incidence than in cell lines derived from other types of tissues. The pRB gene is also lost in many gliomas, but only in high-grade tumors [50] (W. Hamel, unpublished observations). This holds true in cell lines as well as in primary samples, thus making it less likely that the loss is

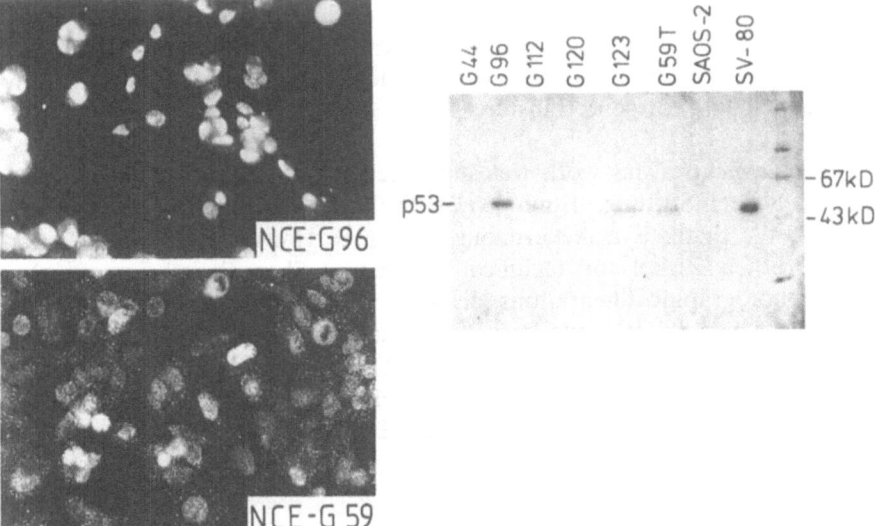

Fig. 8. *Left*: Immunofluorescence of two glioma cell lines after staining with monoclonal antibody PAb1801. The nuclear staining of NCE-G96 (top) which has no mutation in the highly conserved regions of the p53 gene is comparable to the staining of NCE-G59 (below) which bears a point mutation at codon 273 leading to an amino acid exchange. *Right*: Such findings are complemented by immunoblot experiments. Immunoprecipitation was done using PAb1801. The precipitates were separated on a SDS-PAGE gel and transferred to nitrocellulose. The detection was achieved using a rabbit anti-p53 antiserum and a radioactive second antibody against rabbit IgG. Three types of signal were obtained and can be compared with the control cell lines SV-80 (positive control) and SAOS-2 (negative control). Cells showing an intense signal are suspicious of being mutant for p53 and it is tempting to declare cell lines exhibiting weaker signals wild type. In this example both is equally deceptive as mutations were found in NCE-G112, NCE-G123 and NCE-G59 but not in NCE-G96. The negative lanes indeed indicate either that these cells have lost the p53 gene completely or that the molecule is mutated beyond recognition of the antibody (NCE-G44) and (NCE-G120)

due to cell culture adaptation. It thus appears that losses of cellular control or "suppressor" genes are frequent in gliomas and occur independently from each other, as only one cell line (NCE-G130) has lost function of both genes, at least when analyzed on the protein level.

Perspectives

While the description of the growth factor pathways has contributed greatly to our biological understanding of the glioma disease, little has emerged that is of use for therapeutic purposes. The overexpression of EGF-R may provide a target molecule for monoclonal antibody therapies [12], which might even become highly specific when directed against glioma-associated mutants. The antibodies may then be used to shuttle interfering substances to the tumor. We believe that growth factors are epiphenomenal to a superimposed program brought on by an elementary transforming event such as loss of function of a suppressor gene such as p53, RB or the gene yet to be identified on chromosome 10. It will be very tempting to replace defective genes such as the suppressor genes [24] by whatever means may become available for gene transfer.

Acknowledgments. This work was supported by the Deutsche Krebshilfe – Mildred Scheel Stiftung, Bonn (W 41/90/He-2) and the Heinrich Bauer Stiftung. We thank Eva Ackermann, Marianne Hänsel, Hildegard Nausch and Dorothea Zirkel for technical assistance. Sker Freist's preparation of the photographic illustrations deserves special thanks. Fritz Hölzel is thanked for his advice on the karyotypic analyses. The operating room staff are acknowledged for their cooperation. This report contains parts of the doctoral theses of L. Anker and W. Hamel which will be presented to the Fachbereich Medizin at the University of Hamburg.

References

1. Aaronson SA (1991) Growth factors and cancer. Science 254:1146–1153
2. Bigner SH, Burger PC, Wong AJ, Werner MH, Hamilton SR, Muhlbaier LH, Vogelstein B, Bigner DD (1988) Gene amplification in malignant human gliomas: clinical and histopathological aspects. J Neuropath Exp Neurol 47: 191–205
3. Bögler O, Wren D, Barnett SC, Land H, Noble M (1990) Cooperation between two growth factors promotes extended self renewal and inhibits differentiation of oligodendrocyte-type-2 astrocyte (O2A) progenitor cells. Proc Natl Acad Sci USA 87:6368–6372
4. Burger P, Heinz ER, Shibata T, Kleihues P (1988) Topographic anatomy and CT correlations in the untreated glioblastoma multiforme. J Neurosurg 68: 698–704

5. Chellappan SP, Hiebert S, Mudryi M, Horowitz JM, Nevins JR (1991) The E2F transcription factor is a cellular target for the RB protein. Cell 65:1053–1061
6. Deinhardt F (1980) Biology of primate retroviruses. In: Klein G (ed) Viral oncology. Raven, New York, pp 357–398
7. Erickson CA, Turley EA (1987) The effects of epidermal growth factor on neural crest cells in tissue culture. Exp Cell Res 169:267–279
8. Friend SH, Horowitz JM, Gerber MR, Wang XF, Bogenmann E, Li FP, Weinberg RA (1987) Deletions of a DNA sequence in retinoblastomas and mesenchymal tumors: organization of the sequence and its encoded protein. Proc Natl Acad Sci USA 84:9059–9063
9. Fults D, Tippets RH, Thomas GA, Nakamura Y, White R (1989) Loss of heterozygosity for loci on chromosome 17p in human malignant astrocytoma. Cancer Res 49:6572–6577
10. Gross JL, Morrison RS, Eidsvoog K, Herblin WF, Kornblith PL, Dexter DL (1990) Basic fibroblast growth factor: a potential autocrine regulator of human glioma cell growth. J Neurosci Res 27:689–696
11. Haley JD, Hsuan JJ, Waterfield MD (1989) Analysis of mammalian fibroblast transformation by normal and mutated human EGF receptors. Oncogene 4: 273–283
12. Hall WA, Fodstad O (1992) Immunotoxins and the central nervous system. J Neurosurg 76:1–12
13. Haluska FG, Tsujimoto Y, Croce CM (1987) Oncogene activation by chromosome translocation in human malignancy. Annu Rev Genet 21:321–345
14. Heldin CH, Betzholz C, Claesson-Welsh L, Westermark B (1987) Subversion of growth regulatory pathways in malignant transformation. Biochim Biophys Acta 907:219–244
15. Henn W, Blin N, Zang KD (1986) Polysomy of chromosome 7 is correlated with overexpression of the erbB oncogene in human glioblastoma cell lines. Hum Genet 74:104–106
16. Huang SS, Huang JS (1988) Turnover of the platelet-derived growth factor receptor in transformed cells and reversal by suramin. J Biol Chem 263: 12608–12618
17. Humphrey PA, Wong AJ, Gangarosa LM, Archer GE, Lund-Johansen M, Bjerkvig R, Laerum OD, Friedman HS, Bigner DD (1991) Deletion-mutant epidermal growth factor receptors in human glioblastomas. IXth international conference on brain tumor research and therapy, (abstr) ASILOMAR, Oktolar 1991
18. James CD, Carlbom E, Nordenskjold M, Collins P, Cavenee WK (1989) Mitotic recombination of chromosome 17 in astrocytomas. Proc Natl Acad Sci USA 86:2858–2862
19. James CD, He J, Carlbom E, Nordenskjold M, Cavanee WK, Collins VP (1991) Chromosome deletion mapping reveals interferon α and interferon β-1 gene deletions in human glial tumors. Cancer Res 51:1684–1688
20. Kopp R, Pfeiffer A (1990) Suramin alters phosphoinositide synthesis and inhibits growth factor receptor binding in HT-29 cells. Cancer Res 50:6490–6496
21. Lee WH, Shew JW, Hong FD, Sery TW, Donoso LA, Young LJ, Bookstein R, Lee EYHP (1987) The retinoblastoma susceptablity gene encodes a nuclear phosphoprotein associated with DNA-binding activity. Nature 329:642–645
22. Libermann TA, Nusbaum HR, Razon N, Kris R, Lax I, Soreq H, Whittle N, Waterfield MD, Ullrich A, Schlessinger J (1985) Amplification, enhanced expression and possible rearrangement of EGF-receptor gene in primary human brain tumors of glial origin. Nature 313:144–147

23. Lund-Johansen M, Bjerkvig R, Humphrey PA, Bigner SH, Bigner DD, Laerum O-D (1990) Effect of epidermal growth factor on glioma cell growth, migration and invasion in vitro. Cancer Res 50:6039–6044
24. Mercer WE, Shields MT, Amin M, Sauve GJ, Apella E, Romano JW, Ullrich SJ (1990) Negative growth regulation in a glioblastoma tumor cell line that conditionally expresses human wild-type p53. Proc Natl Acad Sci USA 87: 6166–6170
25. Mikkelsen T, Cairncross G, Cavanee WK (1991) Genetics of the malignant progression of astrocytoma. J Cell Biochem 46:3–8
26. Morrison RS (1991) Suppression of basic fibroblast growth factor expression by antisense oligodeoxynucleotides inhibits the growth of transformed human astrocytes. J Biol Chem 266:728–734
27. Nister M, Hammacher A, Mellström K, Siegbahn A, Rönnstrand L, Westermark B, Heldin CH (1988) A glioma-derived PDGF A chain homodimer has different functional activities from a PDGF AB hetrodimer purified from human platelets. Cell 52:791–799
28. Nister M, Libermann TA, Betsholz C, Pettersson M, Claesson-Welsh L, Heldin CH, Schlessinger J, Westermark B (1988) Expression of messenger RNAs for platelet-derived growth factors and transforming growth factor α and their receptors in human malignant glioma cell lines. Cancer Res 48:3910–3918
29. Ohgaki H, Eibl RH, Wiestler OD, Yasargil G, Newcomb EW, Kleihues P (1991) p53 mutations in non-astrocytic human brain tumors. Cancer Res 51:6202–6205
30. Olivier S, Formento P, Fischel JL, Etienne MC, Milano G (1990) Epidermal growth factor receptor expression and suramin cytotoxicity in vitro. Eur J Cancer 26:876–871
40. Raff MC (1989) Glial cell diversification in the rat optic nerve. Science 243:1450–1455
41. Raff MD, Lillien LE, Richardson WD, Burne JF, Noble MD (1988) Platelet-derived growth factor from astrocytes drives the clock that times oligodendrocyte development in culture. Nature 333:562–565
42. Schlegel U, Moots PL, Rosenblum MK, Thaler HT, Furneaux HM (1990) Expression of transforming growth factor alpha in human gliomas. Oncogene 5:1839–1849
43. Simpson Dl, Morrison R, de Vellis J, Herschman HR (1982) Epidermal growth factor binding and mitogenic activity on purified populations of cells from the central nervous system. J Neurosci Res 8:453–462
44. Sjölund M, Thyberg J (1989) Suramin inhibits binding and degradation of platelet-derived growth factor in arterial smooth muscle cells but does not interfere with autocrine stimulation of DNA synthesis. Cell Tissue Res 256: 35–43
45. Stenman G, Rorsman F, Betsholtz C (1988) Sublocalization of the human PDGF A-chain gene to chromosome 7, band q11.23, by in situ hybridization. Exp Cell Res 178:180–184
46. Stern DF, Hare DL, Cecchini MA, Weinberg RA (1987) Construction of a novel oncogene based on synthetic sequences encoding epidermal growth factor. Science 235:321–324
47. Sugawa N, Ekstrand AJ, James CD, Collins VP (1990) Identical splicing of aberrant epidermal growth factor transcripts from amplified rearranged genes in human glioblastomas. Proc Natl Acad Sci USA 87:8602–8606
48. Takahashi JA, Mori H, Fukamoto M, Igarashi K, Jaye M, Oda Y, Kikuchi H, Hatanaka M (1990) Gene expression of fibroblast growth factors in human gliomas and meningiomas: demonstration of cellular source of basic fibroblast

growth factor mRNA and peptide in tumor tissues. Proc Natl Acad Sci USA 87:5710–5714

49. Torp SH, Helseth E, Dalen A, Unsgaard G (1991) Epidermal growth factor expression in human gliomas. Cancer Immunol Immunother 33:61–64

50. Venter DJ, Bevan KL, Ludwig RL, Riley TEW, Jat PS, Thomas DGT, Noble M (1991) Retinoblastoma gene deletions in human glioblastomas. Oncogene 6:445–448

51. Weber W, Gill GN, Spiess J (1984) Production of an epidermal growth factor receptor related protein. Science 224:294–297

52. Weinberg RA (1991) Tumor suppressor genes. Science 254:1138–1146

53. Westermark B, Magnusson A, Heldin CH (1982) Effect of epidermal growth factor on membrane motility and cell locomotion in cultures of human clonal glioma cells. J Neurosci Res 8:491–507

54. Westphal M, Uhlig H, Rohde E, Herrmann HD (1988) EGF-Rezeptoren in menschlichen Gliomen und ihre biologischen Implikationen. In: Bamberg M, Sack H (eds) Therapie primärer Hirntumoren. Zuckschwerdt, Munich, pp 17–21

55. Westphal M, Brunken M, Rohde E, Herrmann HD (1988) Growth factors in cultured human glioma cells: differential effects of FGF, EGF and PGDF. Cancer Lett 38:283–296

56. Westphal M, Nausch H, Herrmann HD (1989) Cyst fluids of malignant human brain tumors contain substances that stimulate the growth of cultured human gliomas of various histological type. Neurosurgery 25:196–201

57. Westphal M, Herrmann HD (1989) Growth factor biology and oncogene activation in human gliomas and their implications for specific therapeutic concepts. Neurosurgery 25:681–694

58. Westphal M, Ackermann E, Hoppe J, Herrmann HD (1991) Receptors for platelet derived growth factor in human glioma cell lines and influence of suramin on cell proliferation. J Neurooncol 11:207–213

59. Westphal M, Nitschke M, Nausch H, Herrmann HD (1992) Cell lines derived from human gliomas activate multiple autocrine pathways simultaneously. Neurosurgery (submitted)

HER-2/Neu/c-ERB-B2 Amplification:
Clinical Relevance in Mammary Carcinoma

W. Jonat, K. Friedrichs, H. Eidtmann, J. Meybohm and S. Singh

Frauenklinik, Universitätskrankenhaus Eppendorf, Martinistraße 52, W-2000
Hamburg 20, FRG

Institut für Humangenetik der Universität Hamburg, Martinistraße 52, W-2000
Hamburg 20

Introduction

The detection of the amplification of the oncogene HER-2/Neu/c-ERB-B2,
henceforth referred to simply as "ERB-B2," is gaining in importance in
the prognosis and therapy of mammary carcinoma. This paper will review
various possible methods of detection at DNA, mRNA and protein levels
from the viewpoint of the clinically active oncologist and assess the value of
such detection in the areas of diagnosis and therapy.

The results of investigations carried out by a joint working group from
the Department of Obstetrics and Gynecology and the Institute of Human
Genetics at the University of Hamburg as well as those of other groups are
taken into account.

Methods of Detection

ERB-B2 can be detected in multiple normal tissues. A number of adenocarcinomas in various organ systems, including the breast, show amplification of this particular oncogene. If amplification is present, it is always frequently associated with overexpression at the mRNA level (van de Vijver et al.
1987; Kraus et al. 1987). At the same time, in the presence of amplification
at the DNA level and overexpression at the mRNA level, the ERB-B2
protein shows an increase (Mori et al. 1987). Numerous molecular biological
and, subsequent to the development of antibodies against the ERB-B2
protein, immunohistochemical methods are today available for assessing the
clinical relevance of ERB-B2 amplification. Blotting procedures, however,
whether Southern, Northern or Western blots, are not suitable for routine
methods from the clinical viewpoint due to the instrumentation required and
the extensive work involved in carrying out individual analyses. Bearing in
mind that analyses carried out for the purpose of establishing a prognosis

Wagener/Neumann (Eds.)
Molecular Diagnostics of Cancer
© Springer-Verlag Berlin Heidelberg 1993

should involve the consequent testing of all tumor tissue, whether primary or metastatic, such methods, which also require substantial quantities of tissue, are not suitable. Consequently, studies presented to date with these techniques have been based on relatively small patient groups with different bases either in respect of the tumour status or the therapy involved. Immunohistochemical methods could well represent a solution to this problem, especially as such techniques can be carried out for patients with long follow-up periods on formalin-fixed mounted tissue specimens.

A significant number of monoclonal and polyclonal antibodies capable of reacting with ERB-B2 protein have been produced (e.g., Gullick et al. 1987; Mori et al. 1987; Musoko et al. 1989). On assessing the analyses carried out in formalin-fixed, paraffin-mounted specimens, ERB-B2 protein staining can be found only in tissue showing ERB-B2 gene amplification. Tumours with a normal number of ERB-B2 gene copies show no staining. In the case of frozen specimens, however, membrane staining can be observed in specimens with normal gene copy characteristics (Slamon et al. 1989). The interpretation of the staining with respect to clinical relevance is thus dependent on the technique employed. Weighing the clinical relevance of the technique presented a major point in the characterization of an acceptable borderline between so-called positive and negative results as required by the clinician.

Besides difficulties due to heterogeneity especially known for breast cancer, the term amplification demonstrates this problem.

ERB-B2 Amplification and Prognosis

Since the first report by Slamon et al. (1987) on the relevance of ERB-B2 amplification for the prognosis of primary mammary carcinoma, numerous additional publications have appeared. The significant interest in this aspect is due to the problems involved in the treatment of primary carcinoma subsequent to surgery.

Based on a meta-analysis of currently more than 100,000 patients with surveillance times up to 10 years, we now know that adjuvant systemic therapy can result in a reduction in mortality and a prolongation of the recurrence-free state.

Figure 1 shows the progress that can currently be achieved through adjuvant systemic therapy. Dependent on the most important prognosis factor in primary nonmetastasizing carcinoma – involvement of the axillary lymph nodes – there is a curative rate of about 70% in node-negative and 30% in node-positive patients after 10 years in those patients not treated systemically. The remaining 30% of node-negative and 70% node-positive patients presumably died within 10 years due to recurrence. If the whole group of node-negative and node-positive patients were to be treated, an effectiveness of adjuvant systemic therapy, whether hormone or chemo-

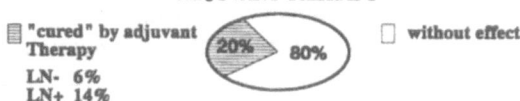

Fig. 1. Present status of adjuvant therapy. *LN*, lymph node

therapy, in one in five cases, i.e., 20% could be assumed. Therapy would be meaningless in 80% of such cases. Thus, a "cure" through adjuvant therapy could, depending on the nodal status, be expected in 6% of node-negative patients and 14% of node-positive patients. Consequently, assuming that the whole patient group were treated, 94 of 100 node-negative patients would have been treated in vain, as they would still have suffered recurrence, or the treatment would have been in vain, as they would have been cured in any case without treatment. Therefore, the aim of the investigation for characterizing the primary tumor is to define the patient group that is to be treated.

Currently it is thought that all node-positive patients should be treated, whereas the situation regarding node-negative patients is somewhat problematic. Therapy currently recommended for both groups is illustrated in Table 1. In characterizing low-risk and high-risk situations, prognosis factors are taken into account; these include morphological criteria such as tumor type and size, various types of receptors – in particular the EGF

Table 1. Consensus statement on the adjuvant therapy of lymph node-negative and lymph node-positive breast cancer. St. Gallen 1992

Node negative	High risk ER −	High risk ER +
Premenopause	CHT	CHT +/− TAM
Postmenopause	CHT +/− TAM	TAM +/− CHT
Senium	TAM (CHT?)	TAM
Node positive	ER −	ER +
Premenopause	CHT	CHT +/− TAM
Postmenopause	CHT +/− TAM	TAM +/− CHT
Senium	CHT (?)	TAM

CHT, chemotherapy; TAM, tamoxifen; ER, estrogen receptor.

receptor which shows a certain homology with the ERB-B2 protein – as well as proliferation parameters like, for example, the S-phase fraction or the ploidy status.

ERB-B2 amplification and overexpression along with other parameters are also taken into account today.

In order to clarify the relevance of ERB-B2 amplification, our group carried out Southern blot analyses on 197 patients with primary mammary carcinoma and compared the results with various other prognosis parameters.

Some 21.5% of the tissue specimens investigated showed ERB-B2 amplification. In respect of nodal status, an amplification of 15.9% was established in the case of node-negative patients in contrast to 25.7% in the case of node-positive patients ($p = 0.259$).

Correlation was also established with the number of affected lymph nodes. Patients with a low axillary tumor burden of one to three lymph nodes exhibited lower amplification than patients with a high number of axillary lymph node metastases. A certain relationship was also established with the receptor status; estrogen receptor-negative patients with unfavorable Prognosis exhibited gene amplification more frequently than their receptor-positive counterparts. This was particularly the case for the progesterone receptor. Histological grading, also a well-known prognosis parameter, showed the highest degree of significance with respect to ERB-B2 amplification. On comparing grade I and II tumors with those of grade III, an amplification of 15% was established in contrast to 27.9% for grade III ($p = 0.085$). No correlation was established with the tumor size, also a prognosis parameter.

On the basis of the examples given, correlation was established between ERB-B2 and other prognosis factors. The Kaplan-Meyer estimate with regard to the free interval is illustrated in Fig. 3. According to this, patients with ERB-B2-amplified tumors have significantly shorter free intervals, based on a monitoring period of 15–52 months (median 30 months). This result is in accordance with other studies.

The most important publications of these on node-negative tumors are summarized in Table 2. Further investigations using larger patient groups undergoing uniform subsequent treatment, and in particular investigations on node-negative patients will be necessary to enable final conclusions to be drawn on the relevance of ERB-B2 amplification in respect of prognosis. Retrospective investigations using immunohistochemistry show, however, that ERB-B2 amplification is relevant in prognosis.

Apart from the relevance in prognosis, there is an interesting correlation between amplification and localization of metastasis. In our 197 patients, a different organ distribution in relation to amplification status was established. As shown in Table 3, visceral metastasis was found more frequently in ERB-B2-positive patients. Liver metastasis was particularly and significantly prevalent. This observation has in the meantime also been confirmed by other groups (Kallioliemi et al. 1991).

Table 2. ERB-B2 amplification in node-negative patients

No relevance in prognosis	
Slamon et al.	(1989)
Tandon	(1989)
Borg	(1991)
Relevance in prognosis	
Tsuda	(1989)
Wright	(1989)
Paik	(1990)
Paterson	(1991)

Table 3. Sites of recurrence by c-erb-B2-amplification status

Site	c-erb-B2 positive ($n = 26$)		c-erb-B2 negative ($n = 42$)		p Value
	n	%	n	%	
Brain	3	11.5	4	9.3	n.s.
Liver	11	42.3	5	11.9	0.02
Lung	9	34.6	11	25.6	n.s.
Other viscera	2	7.7	3	7.0	n.s.
Bone	11	42.3	23	55.8	n.s.
Soft tissue	11	42.3	18	39.1	n.s.
Multiple sites	13	50	21	48.8	n.s.

n.s., not significant.

The prognostic relevance of ERB-B2 amplification is thus possibly not only characterized by increased cell proliferation but also by the prevalence of unfavorable localization of metastasis.

ERB-B2 amplification can hence be employed in deciding whether to use adjuvant therapy or not in the case of primary tumors. However, in cases of metastasis, ERB-B2 amplification currently cannot be employed in deciding whether to use chemotherapy or hormone therapy.

References

Gullick WJ, Berger MS, Bennett PL, Rothbard JB, Waterfield MD (1987) Expression of the c-erbB-2 protein in normal and transformed cells. Int J Cancer 40:246–254

Kallioniemi OP, Holli K, Visakorpi T, Koivula HH, Jorma JI (1991) Association of c-erbB-2 protein over-expression with high rate of cell proliferation, increased risk of visceral metastasis and poor long-term survival in breast cancer. Int J Cancer 49:650–655

Kraus MH, Popescu NC, Amsbaugh SC, King CR (1987) Overexpression of the EGF-receptor related proto-oncogene c-erbB-2 in human mammary tumor cell lines by different molecular mechanisms. EMBO J 6:605–610

Masuko T, Sugahara K, Kozono M, Otsuki S, Akiyama T, Yamamoto T, Toyoshima K, Hashimoto Y (1989) A murine monoclonal antibody that recognizes an extracellular domain of the human c-erbB-2 protooncogene product. Jpn J Cancer Res 80:10–14

Mori S, Akiyama T, Morishita Y, Shimizu S, Sakai K, Sudoh K, Toyoshima K, Yamamoto T (1987) Light and electron microscopical demonstration of c-erbB-2 gene product-like immunoreactivity in human malignant tumors. Virchows Arch [B] 54:8–15

Slamon DJ, Clark GM, Wong SG, Levin WJ, Ullrich A, McGuire WL (1987) Human breast cancer: correlation of relapse and survival with amplification of the HER-2/neu oncogene. Science 235:177–182

Slamon DJ, Godolphin DW, Jones LA, Holt JA, Wong SG, Keith DE, Levin WJ, Stuart SG, Udove J, Ullrich A et al. (1989) Studies of the HER-2/neu proto-oncogene in human breast and ovarian cancer. Science 244:707–712

Styles JM, Harrison S, Gusterson BA, Dean CJ (1990) Rat monoclonal antibodies to the external domain of the product of the c-erbB-2 proto-oncogene. Int J Cancer 45:320–324

van de Vijver M, van de Bersselaar R, Devilee P, Cornelisse C, Peterse J, Nusse R (1987) Amplification of the neu(c-erbB-2) oncogene in human mammary tumors is relatively frequent and is often accompanied by amplification of the linked c-erbA oncogene. Mol Cell Biol 7:2019–2023

Cytokine-Mediated Regulation of Growth Factor Receptors (EGF-R and erb-B2) in Pancreatic Tumors

H. Kalthoff, C. Roeder, and W. Schmiegel

Medizinische Klinik, Universitätskrankenhaus Eppendorf,
Martinistraße 52, W-2000 Hamburg 20, FRG

Introduction

The growth of normal and neoplastic cells is generally believed to depend on a concerted regulation of growth factor signals which are mediated via specific membrane receptors to the nuclear target structures. The family of growth factor receptors is divided into subgroups according to their functional and structural domains (review: [13]). Three very closely related members of the tyrosine kinase subgroup are the epidermal growth factor receptor, *EGF-R/erb-B1* with transforming growth factor alpha (*TGF-α*) and *EGF* as ligands [12], the related *erb-B2/c*-neu receptor with one candidate ligand [10] and the recently descirbed *erb-B3* receptor with no known ligand [8]. Despite the important progress in our knowledge, which has been based on structural analyses after cloning the respective genes, the regulation and the interdependent activities of these signal-transducing cellular compounds remain to be fully characterized. Within the concert of signals triggering cellular growth and differentiation, lymphokines and interleukins – produced predominantly, but not exclusively, by hematopoietic cells – are thought to play a major role. The network of cytokines (review: [1]) comprises factors like tumor necrosis factor alpha (*TNF-α*) and lymphotoxin (*LT* or *TNF-β*) which were originally thought to play a key role in the body's defense against transformed cells, but whose pluripotent activities are now becoming more and more evident [3].

We became interested in the influences of cytokines on epithelial cells when we started to work on the growth regulation of gastrointestinal and particularly pancreatic cancer, aiming to establish a better basis for therapeutic efforts. This paper will present some of our recent data on the expression of the *EGF* receptor (*EGF-R/erb-B1*) and the related *erb-B2/c*-neu receptor in the pancreas and we will further describe the regulation of the corresponding mRNA and protein expression after treating tumor cells with *TNF* or other cytokines as shown in Fig. 1.

Wagener/Neumann (Eds.)
Molecular diagnostics of Cancer
© Springer-Verlag Berlin Heidelberg 1993

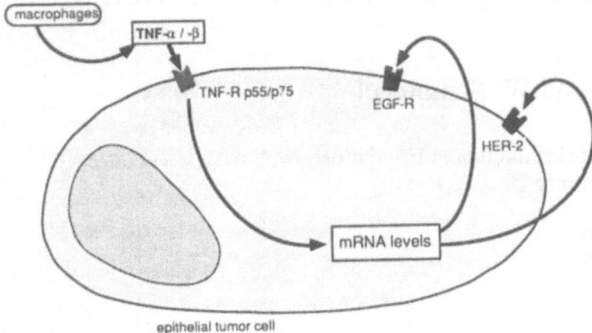

Fig. 1

Results and Discussion

Immunoperoxidase staining of acetone-fixed cryosections from human pancreatic tissues with antibodies specific for *EGF-R/erb-B1* or *erb-B2* respectively showed a high percentage of positivity for both growth factor-receptors in tumor specimen (Table 1). Using three different monoclonal antibodies (mabs) specific for *erb-B2* in parallel, over 50% of pancreatic adenocarcinomas revealed cytoplasmic and/or membrane staining confined to the ductal cells. Based on comparisons with similarly processed breast tumor tissue, which reacted much more strongly, no indication for a high gene amplification rate in pancreatic tumors was found. This finding concurs with the results of Southern blot analyses of nine pancreatic tumor cell lines with one amplified sample only. In an independent set of specimens tested with anti-EGF-receptor mab 425, a comparable range of positive staining was found in the analyzed panel of tumor tissues. No correlation with the differentiation grade of the adenocarcinomas was found, as described previously [7]; however, the proportion of positive samples is

Table 1. Immunoperoxidase staining of pancreatic tissues with erb-B2- and EGF-R-specific monoclonal antibodies

Tissue type	erb-B2/c-neu	erb-B1/EGF-R
Adenocarcinoma	6/11 (54%)	11/21 (52%)
Chronic pancreatitis	1/6 (17%)	5/12 (42%)
Normal pancreas	1/7 (14%)	0/14 (0%)

The three monoclonal antibodies (TA I, CB 11, anti-c-neu/Ab2) which were used for erb-B2 immunostaining showed similar staining patterns and intensities. The respective results are not discriminated here. Staining was restricted to ductal cells. Staining of the EGF-R was achieved by mab 425 (Merck, Darmstadt); only strong staining was scored positive.

likely to be underestimated, because another mab (kindly provided by W. Weber, Hamburg) gave even stronger staining of some frozen sections. Normal pancreatic duct cells failed to show significant immunostaining for either receptor. *EGF*-R and *erb-B2* were also expressed in chronic pancreatitis tissue, albeit less frequently and mostly with lower intensity. The results obtained with nonmalignant tissue specimens suggest that these growth factor receptors may be involved in proliferative responses of ductal epithelial cells to monocytic or lymphocytic infiltration during local inflammation.

In order to establish an in vitro system for studying the regulation of both receptors in response to cytokine treatment of tumor cells, we first decided to investigate a panel of pancreatic tumor cell lines, some of which had been established in our laboratory from primary cultures. The results of the analyses of protein and mRNA expression are summarized in Table 2: 11/12 were positive for *erb-B2* protein, 7/9 for the respective mRNA, with fairly

Table 2. Immunoperoxidase staining of tumor cell lines with erb-B2- and EGF-R-specific monoclonal antibodies and comparison with northern blot analyses

Cell line	erb-B2/c-neu Protein	mRNA	EGF-R/erb-B1 Protein	mRNA
Panc Tu I	+/++	++	+	+/−
Panc Tu II	+/+−	−	+/++	n.d.
QGP-1	+	n.d.	++	n.d.
Colo 357	±	−(++: pA+)	±	+
HPAF	+	+	+	−
Capan 1	+	±(+: pA+)	+	+/−
Capan 2	+	n.d.	++	+
SW 850	−	−	n.d.	+/−
SW 979	+	(±: pA+)	+/±	++
818-1	+/++	(+: pA+)	+	+
818-4	+/++	(+/++: pA+)	+/++	+
818-7	+	n.d.	+	++
PT45-P1	n.d.	(+: pA+)	+	+

The relative intensities of increasing membrane positivity after immunoperoxidase staining are indicated as −, ±, + or ++, with / indicating heterogenous staining. The antibodies used for the receptor protein determination are described in the footnote to Table 1. The results of northern blot analyses with total (20 µg) or poly A+ RNA (pA+) are also given in relative intensities. n.d., not done. The DNA probes used were a 714-bp *Bam*HI/ *Eco*RI human cDNA fragment of the cerb-B2 Amprobe (Amersham, Braunschweig) and a 760-bp *Eco*RI human DNA fragment derived from a A431 cDNA libary (kindly donated by W. Weber, Hamburg). Northern blotting was performed as described previously [5].

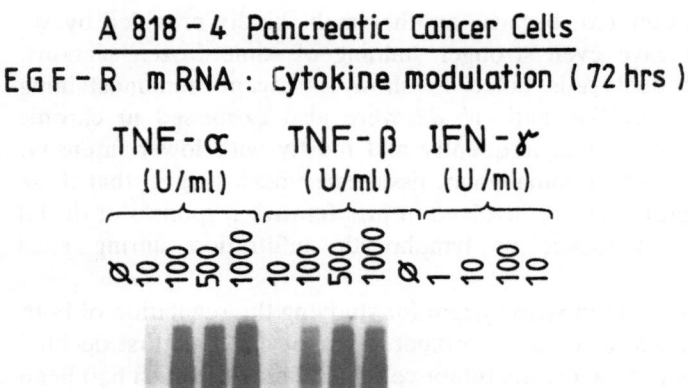

Fig. 2. Expression of EGF-R mRNA after cytokine treatment. Total RNA (20 μg per lane) was subjected to northern blot hybridization. Ethidium bromide staining or hybridization with a constitutively expressed c-raf-1 probe served as control for loading of gels. Transcript sizes (here 10.5 kb) were determined from an RNA ladder (BRL)

good agreement at both levels. In contrast to the *EGF-R* mRNA levels detected by northern blot analyses of 20 μg total RNA, we found polyA$^+$ RNA enrichment to be necessary for the detection of *erb-B2* mRNA in many cases. The expression of *EGF-R* mRNA and protein accompanied the expression of *erb-B2* in the vast majority of cell lines tested. Compared to the overall 50% of receptor-expressing tissue specimens, this finding may indicate an additional in vitro growth advantage and/or selection mode for these two receptor systems.

The pancreatic tumor cell line A818-4 was used for further studies on the regulation of growth factor receptors by cytokines. Northern blot analyses revealed a dose-dependent increase of *EGF-R* mRNA levels in *TNF-α*- or *TNF-β*-treated cells but not in interferon gamma (*IFN-γ*)-treated cells (Fig. 2). The effect of TNF-α was studied on a broader panel of cell lines, extending the observation of *EGF-R* upregulation to all pancreatic cell lines investigated ($n = 6$). Moreover, one of the corresponding ligands, *TGF-α*, was not only constitutively expressed, but also upregulated upon *TNF* treatment in all cell lines (Fig. 3). To extend the findings to the protein level, we investigated the *TNF*-mediated upregulation of the *EGF* receptor by FACS analyses (Fig. 4). The three cell lines studied revealed a significant increase in *EGF-R* binding sites for mab 425 upon cytokine treatment of the cells for 48 h. Further quantitative analyses showed *TNF* to be sufficient for an up to 4.5-fold increase in *EGF* binding sites on A818-4 cells. The mechanism of this *TNF* effect with respect to signal transduction pathways and protein synthesis dependency will be described elsewhere in detail (Schmiegel et al., manuscript in preparation). As many of the cell lines coexpressed *EGF-R* and *erb-B2* mRNA, we wanted to know whether these

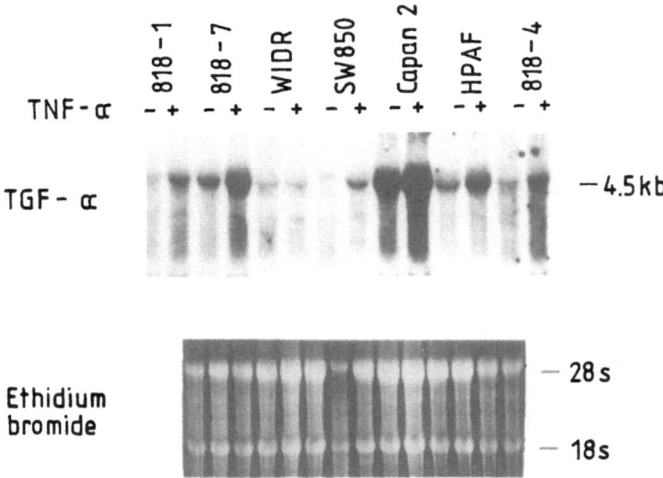

Fig. 3. Expression of TGF-α mRNA after TNF treatment of various tumor cells. RNA was isolated from pancreatic (*WIDR*: colorectal) cancer cells after treatment with 1000 U/ml TNF for 48 h. For northern blotting see legend of Fig. 1

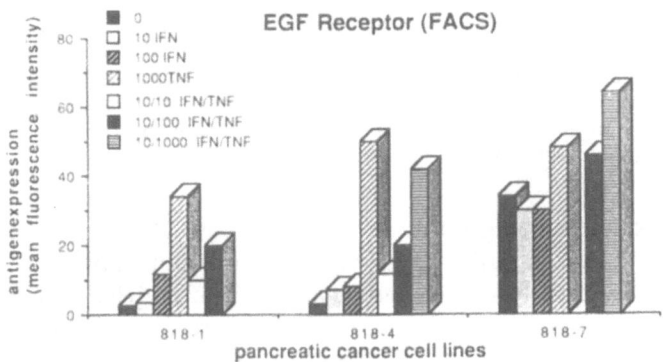

Fig. 4. Expression of EGF-R protein in pancreatic tumor cells after cytokine treatment. Tumor cells were treated with cytokines for 48 h (U/ml as indicated). After mild trypzinisation, cells were incubated with mab 425, followed by indirect FITC staining and FACS analyses

two closely related growth factor receptors are identically regulated via cytokines. To our surprise, a time course experiment with RNA isolated from TNF-treated A818-4 revealed a TNF-induced decrease of *erb-B2* mRNA on northern blots (Fig. 5). Interestingly, this decrease was followed by a reduction in c-myc mRNA levels, suggesting high levels of *erb-B2* to positively influence the growth of these pancreatic tumor cells. As the former northern blot was done with cytoplasmic RNA, we tested a total

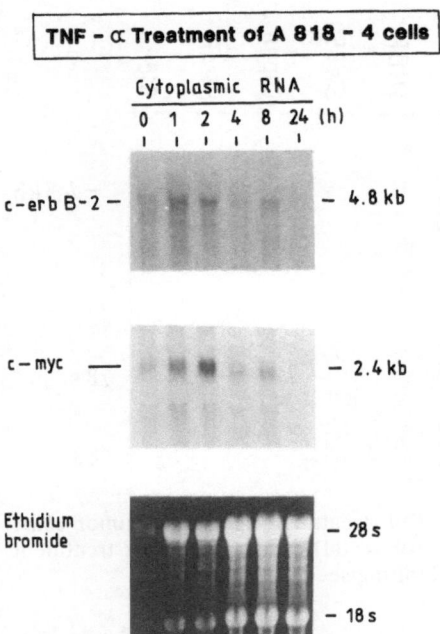

Fig. 5. Expression of erb-B2/c-neu mRNA in TNF-treated tumor cells. Total cytoplasmic RNA was isolated from cells which were grown in the presence of 1000 U/ml for the indicated periods of time. Northern blot was rehybridized with a c-myc probe, indicating a downregulation of this oncogene after a transient (maximum at 2 h) TNF-induced increase (note the reduced amount of RNA from untreated cells in the left lane)

RNA-containing northern blot with both receptor probes in parallel (Fig. 6). Again, a striking inverse relation of both mRNA levels was observed: *EGF-R* mRNA was upregulated 4 h after TNF-incubation of cells, and downregulation of *erb-B2* mRNA started at the same time. Moreover, coincubation of A818-4 cells with TNF and the protein synthesis inhibitor cycloheximide (CHX), revealed that both effects are dependent on de novo protein synthesis, as they could be blocked by CHX. Reprobing of the northern blot with a *c-raf-1* probe served as a control for loading of the gel with 20 µg of total RNA per lane. The striking effects of *TNF* on these two growth factor receptors was confirmed by another independent experiment, where TNF was shown to be superior to *IFN-γ* and a combination of *TNF* with *IFN*; *TNF-α* was substitutable by *TNF-β*. As mentioned above for the *EGF-R* protein, we also studied the levels of *erb-B2*-encoded protein after TNF treatment of pancreatic tumor cells. These studies were performed using a novel enzyme immunoassay and, as shown in Fig. 7, a dose-dependent decrease of *erb-B2* protein in cellular extracts of A818-4 cells was detected 36 h after addition of TNF. Taken together, all assays performed so far undoubtedly unveil an inverse *TNF*-induced regulation of the two closely

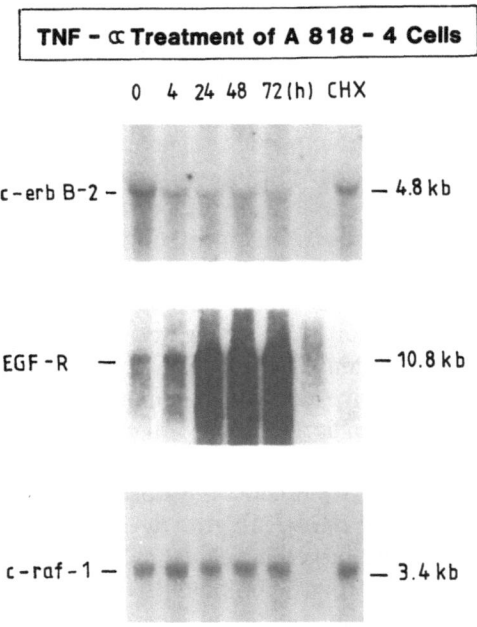

Fig. 6. Inverse regulation of erbB-2- and EGF-R-mRNA by TNF. An amount of 1000 U/ml TNF was used in this time course experiment. RNA in the right lane was isolated from cells which were co-incubated with TNF and cycloheximide (5 µg/ml) for 4 h. Rehybridization with a c-raf-1 probe served as a control for loading the gel, as this mRNA was found not to be regulated by cytokines (in contrast to actin)

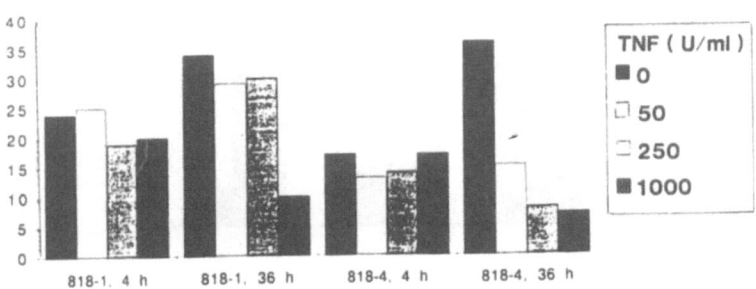

Fig. 7. Decrease of erbB-2 protein in TNF-treated Cells. Pancreatic tumor cell lines A818-1 and A818-4 were treated with increasing doses of TNF for the indicated time. Cellular extracts were prepared according to the mnanufacturer's protocol for the erb-B2/c-neu ELISA (Oncogene Science/Dianova). The relative concentration of erb-B2 protein in U/ml was standardized to milligrams of total protein

related and often coexpressed growth factor receptors, with an increase in *EGF-R* when *erb-B2* is decreased.

From a functional point of view it is of interest that the net effect of TNF treatment of these pancreatic adenocarcinoma cells is a negative one with respect to growth [11]. In general, the *EGF* receptor is believed to positively contribute to proliferation of cells. Consequently, an increase of the respective mRNA and protein, concurred by an increase of the corresponding ligand, should result in a stimulation of growth. The negative outcome in our system points either to a dominant effect of *erb-B2* (due to its decrease) or to additional unknown factor(s) triggered by *TNF*. In an attempt to shed some light on this, we treated A818-4 cells with antisense oligonucleotides directed against the mRNAs coding for *EGF-R*, *erb-B2* or *TGF-a* and with the respective control oligonucleotides. As seen in Fig. 8, treatment with *EGF-R*-specific antisense oligonucleotides (kindly provided by G. Keri, Budapest) reduces the incorporation of thymidine into newly synthesized DNA by A818-4 cells in a dose-dependent manner. This suggest a positive role for the *EGF-R* in growth regulation. Corresponding analyses of *TGF-a*- and *erb-B2*-targeted antisense oligonucleotides also revealed a positive contribution of these compounds towards cellular growth (Fig. 9). In addition, we tested the functional significance of *EGF-R* and other tyrosine kinase receptors by using specific inhibitors previously described as tyrphostins [9]. The results, demonstrated in Fig. 10, again underline the positive contribu-

Treatment of A818-4 Cells with anti-sense Oligonucleotides

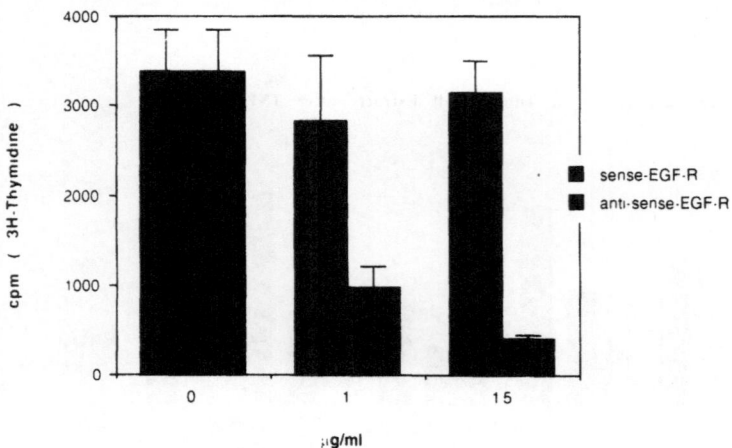

Fig. 8. Decreased DNA synthesis in EGF-R-anti-sense oligonucleotide-treated pancreatic tumor cells. Cells (7×10^4/ml) were grown with 0.5% FCS in microtiter plates and incubated with sense control oligonucleotides or antisense oligonucleotides (kindly donated by G.Keri, Budapest) targeting the EGF-R mRNA. DNA synthesis was labeled with $2\,\mu$Ci/ml ^{3}H-thymidine over the last 4 h of a total incubation period of 24 h prior to harvesting

Treatment of A818-4 Cells with anti-sense Oligonucleotides

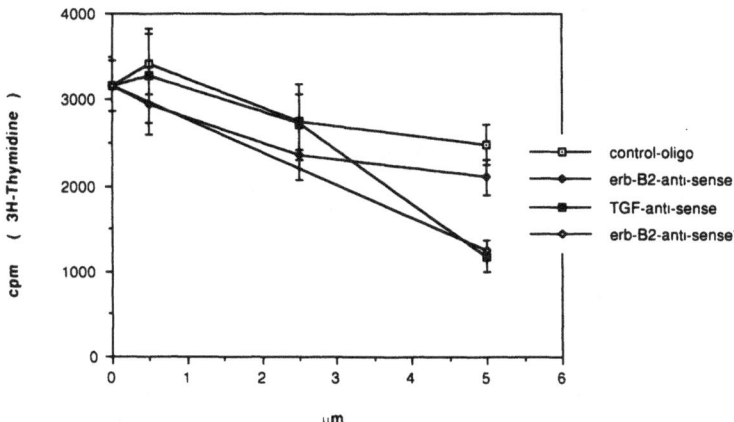

Fig. 9. Influence of erb-B2 and TGF-α antisense oligonucleotides. Pancreatic tumor cells were incubated with increasing concentrations of phosphothioate oligonucleotides (Biometra) for 48 h as described in the legend to Fig. 7. * Another stabilized version of erbB-2 antisense oligonucleotide, prepared by G.Keri, Budapest

Treatment of A818-4 Cells with Tyrphostins

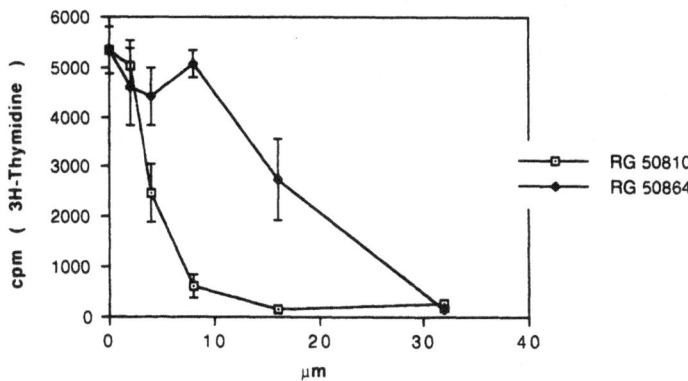

Fig. 10. Inhibition of DNA synthesis in tyrphostin-treated cells. Tumor cells were incubated with tyrosine kinase inhibitors (kindly donated by G. Keri, Budapest) for 27 h in the presence of 0.5% FCS. Cellular growth assay was done as described in the legend to Fig. 7

tion of these receptors by showing a drastic and dose-dependent decrease in thymidine incorporation after treatment of A818-4 cells with two different tyrphostins. However, when we used the *EGF* receptor for an exogenous stimulation of cell growth by the corresponding ligand(s), to our surprise we observed a striking inhibition of proliferation after adding increasing amounts of *EGF*, starting with extremely low doses (Fig. 11). The negative

Treatment of A818-4 Cells with Growth-Factors

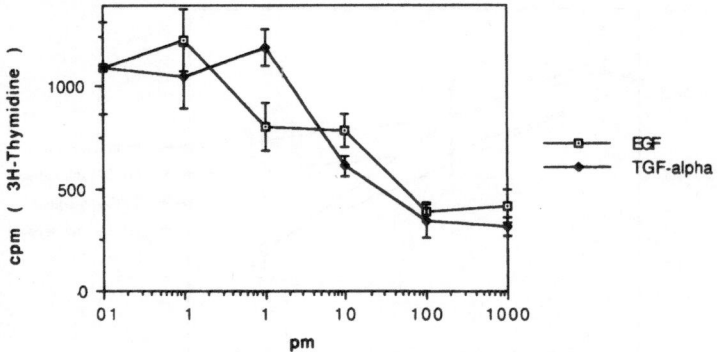

Fig. 11. Long-term incubation of A818-4 tumor cells with growth factors. Monolayer cells were grown in 0.5% FCS and incubated with either epidermal growth factor or TGF-α as indicated. Medium was changed twice during the 1-week treatment. Growth assay was done as described in the legend to Fig. 7

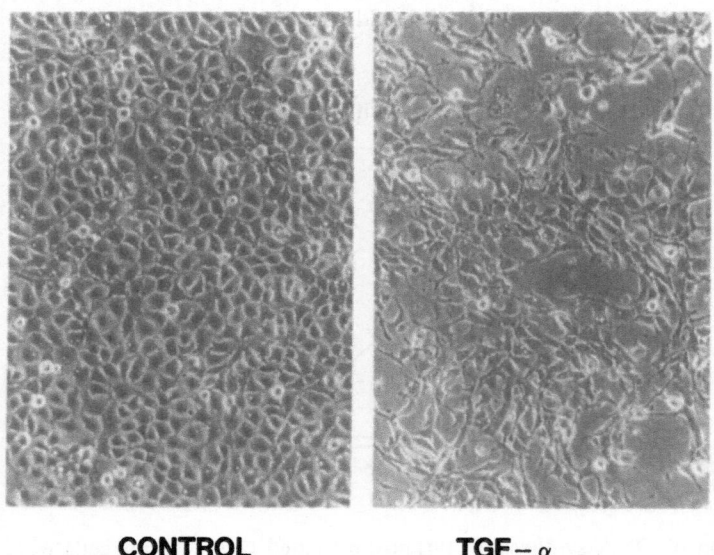

CONTROL   **TGF-α**

Fig. 12. Morphological changes of A818-4 cells induced by *TGF-α*. Monolayer cells were treated with 20 nM TGF-α overnight prior to phase-contrast microscopy

effect of the other ligand, *TGF-α*, was even more dramatic. Neither shorter incubation periods nor pulse treatment inverted the inhibition achieved by both *EGF-R* ligands. The antiproliferative activity of *TGF-α* was accompanied by drastic morphological changes when this growth factor was added to pancreatic tumor cells, indicating the induction of differentiation-related processes (Fig. 12). These changes were even more dramatic than those observed after *TNF* treatment [6].

In summary, the opposing effects of exogenously added growth factors and endogenously synthesized ligands (the positive effects of which are seen when the autocrine loop was either functionally or structurally blocked by tyrphostins or anti-sense oligo's, respectively) can be interpreted by two mechanisms: the first possibility is based on a cooperative effect (cross-talk) between the two receptors, *EGF-R* and *erb-B2*, which is supported by the recent finding that *EGF-R* and *erb-B2* can indeed function as heterodimers [4]. As long as the *erb-B2* ligand is not fully characterized or may target both receptors [10], and as at least one other potential ligand for the EGF receptor has been described [2], the complete understanding of this "two-receptor/four-ligand system" is hampered. The second explanation is based on the assumption that "private" autocrine loops contribute to the un-controlled growth, characteristic for the malignant phenotype, and these cytoplasmic effects may strongly differ from extracellular ("public") autocrine loops. The poor response of A818-4 cells to either anti-*EGF-R*- or anti-*TGFα*-blocking antibodies supports this view, but a complete "private" autocrine loop is supposed to enable tumor cell growth independent of the seeding density, which was not the case with A818-4 cells. Our failure, so far, to block pancreatic tumor cell growth in vitro only by inhibitory antibodies against *EGF-R/TGF-α* may also be related to peculiar steric conditions on the surface of these tumor cells.

The cytokine-mediated regulation of growth factor receptors and their corresponding ligands and the functional studies of these effects with respect to proliferation as described in this paper corroborate the complexity of cytokine/growth factor networks in tumors. The findings may warn us against jumping to conclusions based on either nonmalignant models (such as fibroblasts) or on a particular cell line only (such as A431) for a real understanding of the role of the "best studied" growth factor receptor, the *EGF* receptor, in malignancy.

Acknowledgments. We thank B. Büsing, E. George, J. Gieseking, I. Humburg and D. Jüdes for excellent technical assistance.

References

1. Benton HP (1991) Cytokines and their receptors. Curr Opin Cell Biol 3:171–175
2. Ciardello F, Kim N, Saeki T, Dono R, Persico MG, Plowman GD, Garrigues J, Radke S, Todaro GJ, Salomon DS (1991) Differential expression of epidermal growth factor-related proteins in human colorectal tumors. Proc Natl Acad Sci USA 88:7792–7796
3. Fiers W (1991) Tumor necrosis factor: characterization at the molecular, cellular and in vivo level. FEBS Lett 285/2:199–212
4. Goldman R, Levy RB, Peles E, Yarden Y (1990) Heterodimerization of the erbB-1 and erbB-2 receptors in human breast carcinoma cells: a mechanism for receptor transregulation. Biochemistry 29:11024–11028

5. Kalthoff H, Roeder C, Humburg I, Thiele H-G, Greten H, Schmiegel W (1991) Modulation of platelet-derived growth factor A- and B-chain/c-sis mRNA by tumor necrosis factor and other agents in adenocarcinoma cells. Oncogene 6:1015–1021
6. Kalthoff H, Roeder C, Schmiegel W (1992) Expression and regulation of erbB-1 and erbB-2 proto-oncogenes in human pancreatic adenocarcinoma cells. In: Löhning T, Jonat W (eds) International symposium on the clinical and scientific relevance of HER-2/NEU/ERB-B2. Springer, Berlin Heidelberg New York (in press)
7. Klöppel G, Maillet B, Schewe K, Kalthoff H, Schmiegel WH (1989) Immunocytochemical detection of epidermal growth factor receptors (EGFR) and transferrin receptors (TR) on normal, inflamed and neoplastic tissue. Pancreas 4:649 (abstr)
8. Kraus MH, Issing W, Miki T, Popescu NC, Aaronson SA (1989) Isolation and characterization of ERBB3, a third member of the ERBB/epidermal growth factor receptor family: evidence for overexpression in a subset of human mammary tumors. Proc Natl Acad Sci USA 86:9193–9197
9. Levitzki A (1990) Tyrphostin: potential antiproliferative agents and novel molecular tools. Biochem Pharmacol 40:913–918
10. Lupu R, Colomer R, Zugmaier G, Sarup J, Shepard M, Slamon D, Lippman ME (1990) Direct interaction of a ligand for the erbB2 oncogene product with the EGF receptor and p185^{erbB2}. Science 249:1552–1555
11. Schmiegel WH, Caesar J, Kalthoff H, Greten H, Schreiber HW, Thiele H-G (1988) Antiproliferative effects exerted by recombinant human tumor necrosis factor-α (TNF-α) and interferon-γ (IFN-γ) on human pancreatic tumor cell lines. Pancreas 3:180–188
12. Waterfield MD (1989) Epidermal growth factor and related molecules. Lancet 6:1243–1246
13. Yarden Y, Ullrich A (1988) Growth factor receptor tyrosine kinases. Annu Rev Biochem 57:443–478

Chimeric CEA-Specific Antibody Fragments and Anti-idiotypic Antibodies for Immunoscintigraphy and New Therapeutic Approaches to Colorectal Carcinomas

M. Neumaier

Abteilung für Klinishe Chemie, Medizinische Klinik, Universitätskrankenhaus Eppendorf, Martinistraße 52, W-2000 Hamburg 20, FRG

This contribution is intended to illustrate a system of specific in vivo immune diagnosis and therapy of solid tumors and is structured into three sections, each introducing one part of an idiotypic cascade that characterizes this system. The first section gives a personal interpretation of the problem of therapeutic management in the context of colorectal carcinomas, followed by the approach used to construct chimerized immunoglobulin derivatives for tumor targeting and specific passive immunotherapy. Finally, the characterization of an idiotypic network cascade will be described and its implications for active immunotherapy of CEA-expressing tumors will be discussed.

Current Status of Management of Colorectal Cancer

Colorectal cancer is rated among the most common solid carcinomas in western societies with respect to both incidence and mortality. Usually, colorectal carcinomas are diagnosed only at a late stage, and a high proportion of patients already have regional and/or distant metastases. Apart from surgery there are no effective therapeutic measures. Specifically, chemotherapy has proven almost completely unsuccessful.

The UICC meeting in Hamburg in 1990 reached a sobering conclusion concerning the effectiveness of chemotherapy in the therapy of solid tumors. In public statements even the protagonists of combined chemotherapeutic protocols showed increasing ambivalence when assessing the relation between the benefit for the patient and the massive side effects. Similar conclusions have recently been reached by the Arbeitsgemeinschaft für Internistische Onkologie der Deutschen Krebsgesellschaft (AIO). In their compilation of studies (reviewed in [42]) it is stated that the clinical response rates to cytostatic drugs such as 5-fluorouracil and other fluoropyrimidines, mitomycin C and others are below 20%. Combination of drugs increased cytotoxic effects without significant prolongation of survival, and was

Wagener/Neumann (Eds.)
Molecular Diagnostics of Cancer
© Springer-Verlag Berlin Heidelberg 1993

therefore not recommended. Local chemotherapy of, for example, liver metastases of colorectal carcinomas by intra-arterial administration via the proper hepatic artery was not generally recommended. Apart from being strenuous for the patient, the effort involved in this method is immense in relation to the prospects of success. In addition, it simply cannot be regarded as a general therapy concept. Finally, adjuvant chemotherapy for gastrointestinal carcinomas was not recommended by the AIO.

General Characteristics of Colon Cancer

The picture sketched above is particularly unsatisfactory in light of the fact that colorectal carcinomas regularly possess features that set them apart from other gastrointestinal tumors. I believe that these features are worth mentioning, if only to explain, in part, the results of chemotherapy. Systemic chemotherapeutic protocols may fail because colorectal carcinomas, like cancers of other organs (e.g., breast and lung), possess a high proportion of resting cells, in the G_o phase of the cell cycle. In a recent review by Tannock, some general conclusions were reached from studies on the growth rates of tumors of different organ systems [52]. Briefly, tumors that were responsive to chemotherapy (e.g., lymphomas, testicular cancer and tumors in children) were usually growing more rapidly, with doubling times of 4–7 weeks. In the group of less responsive cancers it was found that primary colon carcinomas were by far the slowest growing tumors, with a volume doubling time of 90 weeks. Finally, metastases derived from slowly growing tumors tended to proliferate more rapidly than the tumors from which they were seeded. Due to the anatomic and functional design of the colon, even fair-sized cancers often remain silent for years. Therefore, a large colon tumor is the result of lack of clinical symptoms and not of rapid proliferation. If the rate of proliferation is one criterion for susceptibility of a tumor to chemotherapy, colon cancer is probably a very poor candidate. As a consequence, it may turn out that the slow growth rate of these cancers will severely limit the potential of modern, more biologically oriented therapeutic concepts aiming at the interruption of proposed growth control mechanisms like autocrine and paracrine loops.

On the other hand, diagnostic and therapeutic approaches that reach resting malignant cells in this type of solid, slow-growing tumor would be largely independent of proliferation and could be envisioned as complementing primary therapy, provided these cells could be specifically detected.

The Carcinoembryonic Antigen as Tumor Marker

The human carcinoembryonic antigen (CEA) was first described in 1965 [19] and has since become the best-characterized tumor-associated antigen. An excellent discussion of CEA and its cross-reacting antigens is provided by Zimmermann elsewhere in this volume. Therefore, I would merely like to add here that expression of CEA is maintained in over 90% of all colon tumors [49]. With the exception of its luminal apical expression in the colonic mucosa, CEA is expressed nowhere in the body and ectopic CEA-positive cells can be regarded as malignant. It appears that CEA may, for these reasons, make an excellent target for tumor imaging and specific immunotherapy approaches by CEA-specific antibodies. If so, there are basically three requirements for an antibody to be used in in vivo applications:

1. It has been known for a long time that in addition to CEA there are a number of immunologically related but physicochemically distinct antigens present in different normal tissues and body fluids [2, 15, 26, 35–37, 51]. From molecular cloning of the genes coding for these antigens [1, 22, 38, 41] the CEA gene family has emerged. (For details, see the chapter by Zimmermann). The close relatedness among the members of this gene family emphasizes the critical role of specificity of anti-CEA antibodies that are planned to be used in humans. For example, the administration of antibodies that bind to nonspecific cross-reacting antigen (NCA), a major CEA-related antigen expressed on the surface of granulocytes and macrophages, has led to severe granulocytopenia and side effects in patients [14]. However, there are a number of monoclonal antibodies (mabs) that recognize CEA-specific epitopes. We used the murine mab T84.66, because it possesses one of the highest affinities to CEA (2.6 × 10^{-10} l/mol) [55] and was shown not to cross-react with CEA-related antigens of different sources by means of western blotting and enzyme-linked immunosorbent assay (ELISA) [36, 37, 56]. Comparison of immune reaction of mab T84.66 with a cross-reactive antibody is shown in Figs. 1 and 2. Immunohistochemistry data demonstrate that mab T84.66 only shows a weak reaction with epithelial structures of pancreas acini [35]. It has been shown that the ^{111}in-labeled murine mab T84.66 is well suited for tumor targeting both in animal xenograft models [23] and in humans [4]. Typical results obtained with T84.66 are shown in Figs. 3 and 4. Clinical studies have been performed on more than 200 patients so far (A. Raubitschek and J.E. Shively, personal communication).

2. The use of xenogeneic immunoglobulins in humans almost always results in "Human anti-mouse antibodies" (HAMA) [13, 28, 57]. Since cancer patients need long-term monitoring, this immune response clearly limits the necessary repeated in vivo application of antibodies. Although serious clinical complications with immediate-type allergic reactions are rarely observed, the formation of immune complexes between HAMA and the xenogeneic immunoglobulin drastically reduces the antibody's half-life in the circulation, therefore rendering it inefficient for the intended purpose. Most frequently, the HAMA response is directed against the Fc part of the xenogeneic mab that is coded for by the

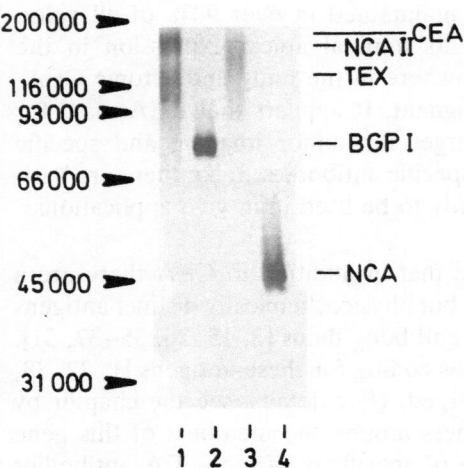

Fig. 1. Binding of mab T84.1-E3 to CEA and related antigens from perchloric acid extracts in Western Blot analysis. Lane 1: primary colorectal adenocarcinoma. Lane 2: normal human bile. Lane 3: normal meconium. Lane 4: human spleen. T84.1-E3 was used at a concentration of 0.51 mg/L

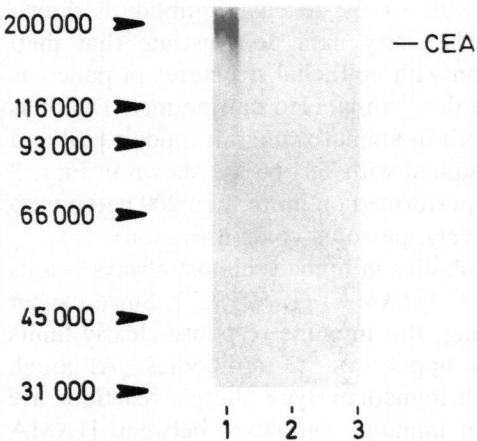

Fig. 2. Binding of mab T84.66 A3.1-H11 (referred to in the text as T84.66) to CEA and related antigens from perchloric acid extracts in Western Blot analysis. Lane 1: primary colorectal adenocarcinoma. Lane 2: normal human bile. Lane 3: normal human spleen. T84.66 was used at a concentration of 0.51 mg/L

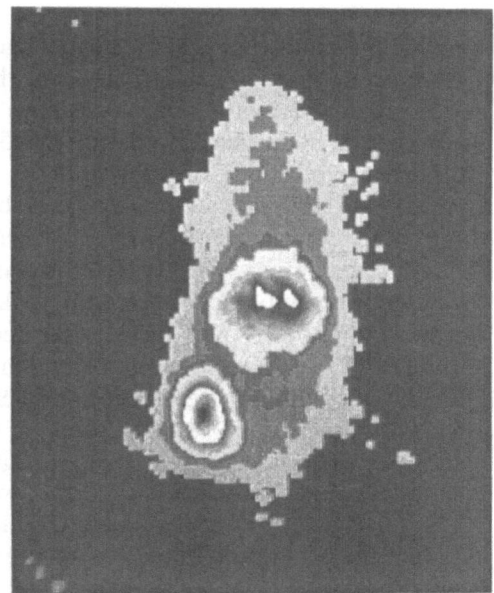

Fig. 3. Radioimmunoszintigraphy of a nude mouse bearing a colorectal carcinoma transplant in the hind leg. Antibody used was ^{111}In-labelled murine T84.66. Note the high liver uptake of the antibody. (By kind permission of Drs. J.D. Beatty and J.E. Shively, City of Hope Natl. Medical Center, Duarte, California)

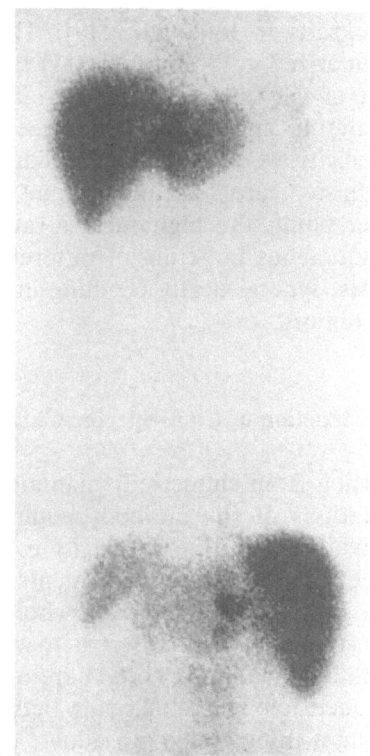

Fig. 4. Radioimmunoszintigraphy of liver metastasis in a patient with colorectal cancer. Top part: anterior view. Bottom part posterior view. Two metastases are clearly discernable despite the high unspecific liver uptake. Metastases have been confirmed surgically (By kind permission of Drs. J.D. Beatty and J.E. Shively, City of Hope Natl. Medical Center, Duarte, California)

constant region genes of light and heavy chain. Chimeric human/mouse immunoglobulins, in which the xenogeneic constant region genes are replaced by human counterparts, are thought to be the solution to this problem, and a number of groups have published reports on the generation of such antibodies [5, 30, 40, 41, 45, 50]. Maximum humanization of xenogeneic antibodies is achieved by transplanting the hypervariable regions (CDRs) of the xenogeneic mab onto the backbone of a human antibody [44, 54]. However, since certain framework residues in the variable region genes are crucial for conformation of CDRs [11, 12], this loop-grafting approach can result in a major loss of affinity of these grafted antibodies. We have generated chimeric anti-CEA antibodies that retain the complete variable region genes.

3. Finally, the antibody's efficiency in reaching the tumor target will depend on a number of parameters, one of which clearly is the size of the antibody. The superiority of immunoglobulin fragments over complete antibodies is a well-documented fact with xenogeneic antibodies [8–10, 20]. Since genetic engineering allows this parameter to be changed almost at will, the generation of recombinant antibody fragments of various sizes can be included when designing a chimeric antibody.

In recent years very encouraging results have been reported by the group of Mach and Buchegger on the passive immune therapy of colon tumor xenografts in nude mice [9]. The importance of their approach can be summarized in three points: (1) the tumors were well established before the start of the treatment; (2) the authors used ^{131}I-labeled F(ab')$_2$ fragments of murine antibodies, and three antibody fragments of different epitope specificity were used simultaneously; (3) the treated mice were given full "intensive care," including bone marrow transplantation. Disregarding the latter point, the high success rate of this model can be explained by the simultaneous targeting of different epitopes together with the use of fragments, synergistically resulting in the delivery of a higher radiation dose to the tumor.

Construction of CEA-Specific Chimeric Human/Mouse Antibody Fragments

Starting from chimeric human/mouse T84.66 [39], we have generated gene mutations of the antibody coding for different chimeric immunoglobulin derivatives with the specificity of T84.66 as a systematic approach to study the effects of different fragments with the same antigen-binding properties. The genes constructed from whole chimeric T84.66 are shown in Fig. 5. For the constructs a number of manipulations of the human hinge region of immunoglobulin G1 (IgG1) are necessary. The hinge region of human IgG1 is coded for by a small exon between antibody domains CH$_1$ and CH$_2$ and contains three cysteine residues. The first cysteine forms the disulfide bridge

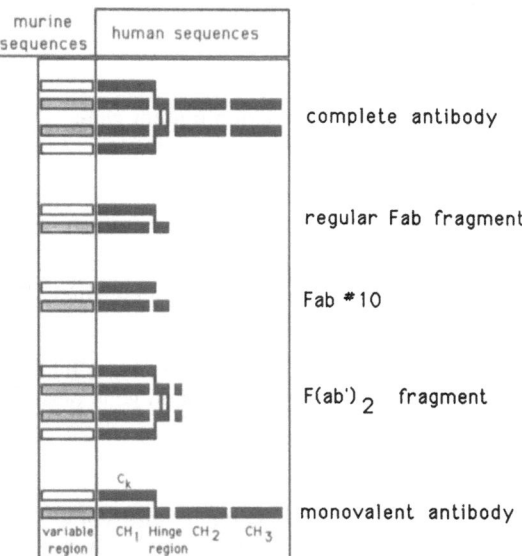

Fig. 5

with the respective residue of the light chain, while the other two cysteines create disulfide bonds between the heavy chains. Specifically, two Fab fragments were generated that differed by the presence or absence of the first hinge region cysteine. Therefore, the chains in the corresponding fragment are held together only by hydrophobic interactions. The gene construct for the $F(ab')_2$ was generated by introducing a reading frame shift just 3' of the most C-terminal cysteine residue. This led to a translational stop codon in the next antibody domain CH_2. The monovalent antibody fragment was generated by changing cysteine codons 2 and 3 to glycine codons, thereby eliminating the disulfide bridges between the heavy chains. The expected molecular weights for Fab, monovalent fragment and $F(ab')_2$ are around 50 kDa, 75 kDa and 100 kDa respectively.

The site-directed mutagenesis methods we employed included plasmid-based mutagenesis protocols with slight modifications that allowed for selection of mutations over wild-type hinge region genes [27, 47]. These methods generally gave only very low mutation frequencies caused by the size of the genomic inserts of our chimeric constructs. We therefore developed a mutagenesis method which is shown schematically in Fig. 6. The first step includes the generation of a hinge deletion mutant (Fig. 7). It has to be noted that two of the suitable restriction sites are dcm-methylated in plasmid DNA prepared from normal *E. coli* K12 bacteria. Therefore the DNA was prepared from a dcm⁻ strain previously transformed with the plasmid. This allows the removal of the hinge domain after partial digestion with the restriction enzyme *Stu*I. Since in the deletion mutant plasmid a *Stu*I site is regenerated that cannot be methylated compared to the site at position 21 of

wild type human IgG 1
constant region gene

HINGE deletion mutant gene
with unique cloning site

generation of a wild type HINGE region gene
cassette by enzymatic DNA amplification (PCR)

site-directed mutagenesis of cassette by PCR
for generation of HINGE region gene derivatives

cloning into HINGE region deletion mutant
characterization of new gene by sequencing

expression in eukaryotic cells producing
the chimeric immunoglobulin light chain

Fig. 6. General strategy for the construction of chimeric human/mouse immunoglobulin derivatives by manipulation of the human HINGE region gene

the gene (Fig. 7), growth of the plasmid in dcm⁺ bacteria yields DNA that can only be restricted with *Stu*I at the site of the deletion. The mutated hinge regions described below were inserted in this cloning site to yield genes coding for the respective antibody fragments.

The amplification of the hinge region by the polymerase chain reaction (PCR) [46] is shown in Fig. 8. Briefly, oligonucleotide primers containing internal *Eco*RI restriction sites were used to amplify the region from the wild-type gene, followed by restriction with *Eco*RI. Ligation of the restricted DNA was carried out under conditions favoring monomeric circles [31]. A second round of PCR was performed using primers, one of which harbored the mutation. This amplification can only be successful if the DNA fragment obtained during the first round is ligated to circles or head-to-tail concatemers. After this second PCR, the amplified product was ligated under the same conditions as above, followed by restriction with *Eco*RI, which reconstitutes the proper arrangement of the cassette. Since one of the primers in this second PCR represents the mutagenic oligonucleotide, all fragments contain the intended mutation by definition. The mutated hinge region cassette was then inserted into the *Stu*I site of the hinge deletion plasmid by blunt end ligation. Since this regenerated the *Eco*RI sites at both ends of the cassette, a convenient cloning site for insertion of other muta-

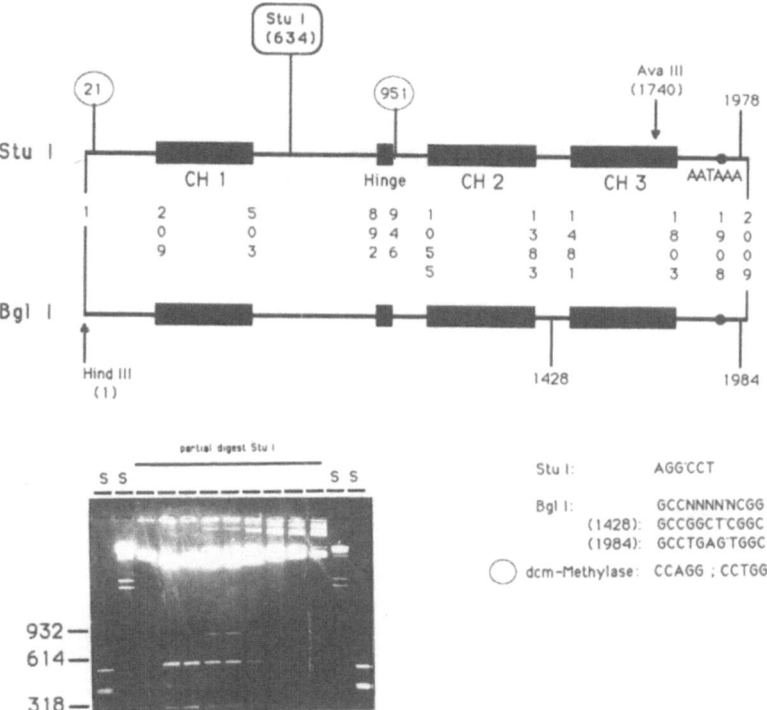

Fig. 7. Restriction map of the human IgG1 constant region gene for the restriction enzymes Stu I and Bgl I. Domain exons of the human IgG1 gene are depicted by black boxes. Stu I restriction sites were used for the generation of the deletion mutant. Recognition sequences for are given in the lower right part of the figure. Methylated sites are indicated. Boxed Stu I site at position 634 has not been reported. The deletion mutant was obtained after partial digestion with Stu I as shown in the left lower part of the figure. S: Molecular weight standard. Numbers refer to respective Stu I fragments of human the HINGE region gene

tions or the creation of fusion proteins was obtained. DNA sequence analysis showed that all the plasmids obtained with this rapid system harbored the intended mutations. All chimeric antibody fragments shown in Fig. 5 have been expressed in eukaryotic cells after electroporation as described previously [39]. In this study whole chimeric T84.66 has been shown to have antigen-binding properties identical to the murine mab. As expected, chimeric antibody fragments secreted by cell lines previously transfected with the respective gene constructs also show the anti-CEA specificity of the parent T84.66. As an example of comparable binding of CEA, whole chimeric T84.66 and a chimeric Fab fragment are shown in an inhibition ELISA (Fig. 9). The fragment used (clone Fab 10) resembles the chimeric T84.66 fragment lacking the disulfide bond between light and heavy chain and therefore was regarded as the most unstable construct among the muta-

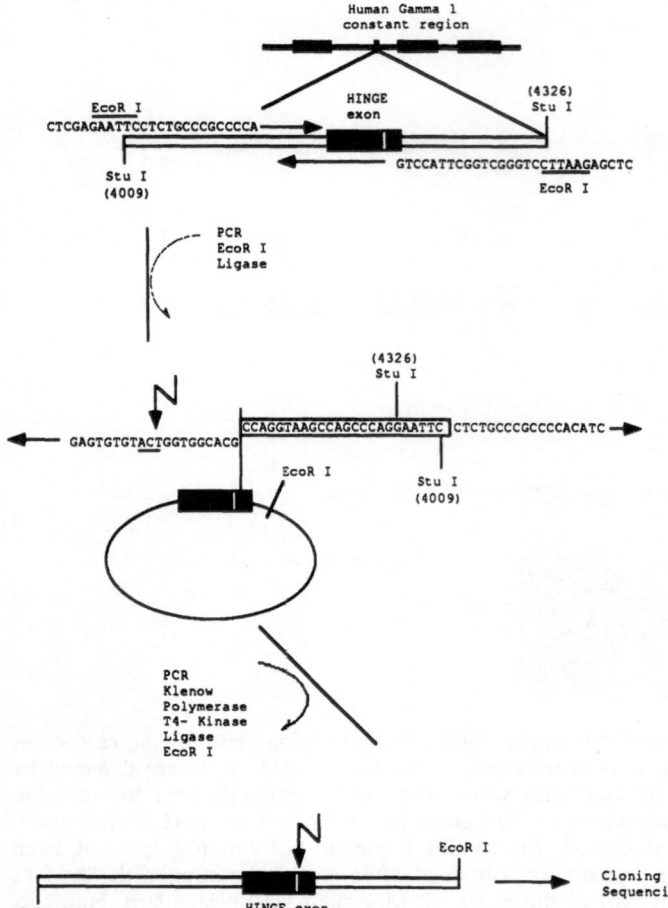

Fig. 8. PCR-based site-directed mutagenesis of the human HINGE region gene. Primer sequences for the first round of amplification contain internal restriction sites (Eco RI). Mutagenic primer for the second round of amplification is indicated by lightning bolt with the mutation underlined. The other primer for the second amplification is boxed

tions generated. Binding of whole chimeric antibody (clone 6IC1) to immobilized CEA was inhibited using serial dilutions of biotinylated murine T84.66. Superimposable inhibition curves were obtained for the chimeric antibody molecules. This indicates that the Fab fragment possesses an affinity constant for CEA comparable to that of the murine T84.66 immunoglobulin. By ELISA the existence of a human Fc terminus was shown by an Fc-specific polyclonal goat antiserum for the monovalent fragment (Dianova, Hamburg), while it was missing in the F(ab')₂ fragment and in both Fab fragment mutations. Further thorough characterizations that precisely assess

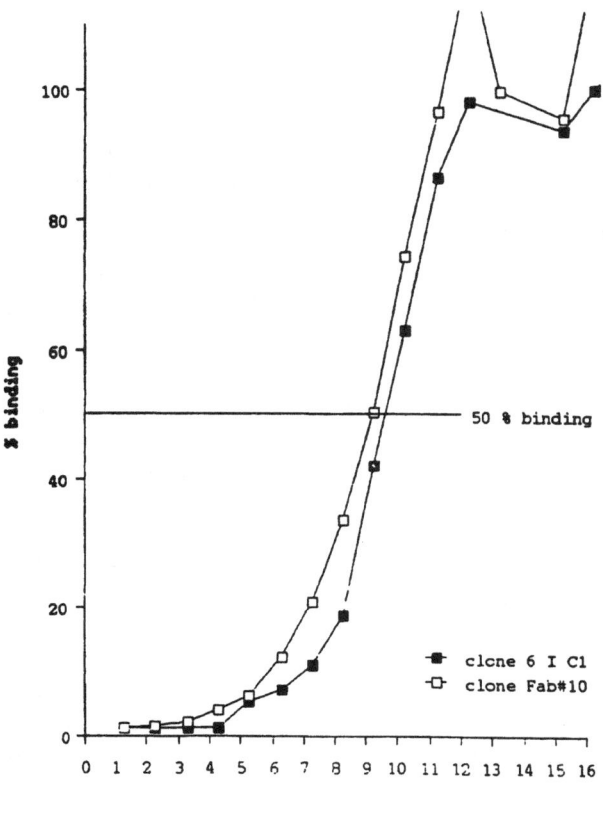

Fig. 9. Comparison of binding of chimeric immunoglobulin molecules to CEA. Clone 6 I C1: whole chimeric T84.66 immunoglobulin. Clone Fab#10: chimeric T84.66 Fab fragment lacking the inter-chain disulfide bond (see also Fig. 5)

both affinity constants and molecular weights will be published elsewhere (Neumaier et al. manuscript in preparation).

We hope that in vivo experiments using these chimeric antibody fragments will provide us with the opportunity to study biodistribution, clearance profiles and tumor penetration using molecules of the same specificity and affinity but different sizes. This will be especially interesting for the monovalent antibody fragment because of its favorable intermediate size. The possible mediation of biological effector functions may make this construct a useful tool for both imaging and therapy.

Generation of Monoclonal Anti-idiotypic Antibodies Mimicking
a CEA-Specific Epitope

In immunoglobulins, the binding of an epitope and the mimicry of this same epitope represent two opposite principles of one single structural concept, its "negative" and "positive" form respectively. At the interface between these interacting complementary forms the specificity and function of antibodies is decided.

As a concept for active specific immunotherapy, anti-idiotypic antibodies have been suggested to trigger the patient's immune system against his tumor burden, an idea based on the idiotype network theory proposed by Jerne in 1974 [24]. This theory states that one can generate, through an immunological cascade of network interactions, antibodies which would represent the internal image of given epitopes of a nominal antigen. Anti-idiotypic immunoglobulins could act as network antigens and induce humoral and cellular cytotoxic effects against antigen-bearing cells. The usefulness of such antibodies as vaccines has been shown for a number of bacterial and viral antigens (reviewed in [48]). Promising antitumor effects of idiotype vaccines in experimental animal systems have been described [16, 28, 43]. With respect to human studies, anti-idiotypic antibodies have been reported for the treatment of solid tumors such as melanomas [33] and gastrointestinal carcinomas [21]. In the latter study, Herlyn et al. used a goat polyclonal anti-idiotype vaccine and reported partial clinical remissions in some colon cancer patients. The majority of their patients showed an increased humoral response towards the nominal antigen after therapy, demonstrating that some antibodies in the anti-idiotypic goat serum provoked an antigen-specific anti-anti-idiotypic response (Ab_3). This may indicate that idiotype vaccines bearing internal images of human tumor-associated antigens have potential as modulators of the immune response in malignant diseases. In this context, it has been demonstrated that the presentation of epitopes might, in the context of immune network control, break the tolerance to an antigen-positive malignant cell [7]. On the other hand, the induction of humoral antitumor response alone may not be sufficient, and cellular immune mechanisms may be required for major effects in the course of the disease. Unfortunately, no appropriate animal model systems exist that can be used to examine T-cell cytotoxicity functions triggered by a given anti-idiotype, unless antigenic determinants are shared between animal and human tumors. As an example, Mittelman et al. have reported encouraging results after treatment of melanoma patients with a murine monoclonal anti-idiotype (maId) bearing the internal image of a shared determinant on the "high-molecular-weight melanoma-associated antigen" (HMW-MAA) [33]. Unfortunately, such determinants are not available in the CEA system. Until suitable animal systems become available, anti-idiotypes presenting CEA-specific determinants can only be tested for their functional effects in clinical trials. Recently it was demonstrated for the first time that the response against internal images of an antigen can result in delayed-type hypersensitivity reactions in tumor patients [32]. This establishes that anti-idiotypes also trigger cellular immunity functions in tumor rejection.

For reasons laid out above, mab T84.66 was thought to represent an excellent candidate to generate maIds mimicking a CEA-specific antigenic determinant. It was speculated that the specificity of the idiotype for CEA may be critical in that harmful autoimmune disease could result from using internal image maIds against mabs binding to CEA-related antigens. Similarly, Bhattacharya-Chatterjee et al. have argued that immunization with complete CEA may provoke autoimmune reactions [3]. Mab T84.66 has been used as the immunogen to generate maIds in syngeneic Balb/c mice [17]. Among a number of hybridoma antibodies that showed binding to paratope-related idiotope structures on T84.66, one (maId 6G6.C4) appeared to functionally mimic the T84.66 epitope according to the following criteria:

MaId 6G6.C4 binds to private idiotopes on mab T84.66, since it reacted neither with murine isotype control immunoglobulins nor with a panel of murine monoclonal anti-CEA idiotypes. Among these is mab CEA.66, which recognizes two epitopes on CEA, one of which has been shown to overlap with the T84.66 epitope [18]. The failure of maId 6G6.C4 to bind to mab CEA.66 can be interpreted in different ways. It is possible that the two epitopes are structurally entirely independent and the observed cross-inhibition is due to steric hindrance. Alternatively, mabs T84.66 and CEA 66 may share parts of their epitopes. In this case, the result indicates either that maId 6G6.C4 does not contain the part of the T84.66 epitope recognized by mab CEA.66 or that it is an accurate image of the T84.66 epitope with the "CEA.66 part" alone not being sufficient to bind to mab CEA.66. In experiments comparable to the one described by Monestier et al. [34], we found that maId 6G6.C4 bound neither the denatured nor the reduced idiotype, indicating that the correctly assembled antigen-binding region of native T84.66 is required for the formation of the immune complex. In addition to immunochemical characterization, maId 6G6.C4 can be used to raise CEA-specific polyclonal Ab_3 antisera in rabbits. This strongly suggests that the maId functionally mimics CEA in vivo. Specifically, in ELISAs the CEA/Ab_3 immune complex can be inhibited to background values by addition of soluble CEA. In western blots, Ab_3 antiserum specifically recognizes CEA of different sources (Fig. 10). No reaction was observed with up to 100 µg of a crude lung tumor extract containing NCA, a major CEA-related glycoprotein present, for example, on the surface of granulocytes, or with up to 100 ng of purified NCA (Fig. 11). When diluted to give comparable staining, a commercially available anti-CEA antiserum clearly detected purified NCA and NCA in the tumor extract. These results clearly demonstrated that the Ab_3 response against the anti-idiotype had a high specificity for CEA in that it did not recognize epitopes shared between CEA and NCA. Two CEA maIds have recently been reported by others in

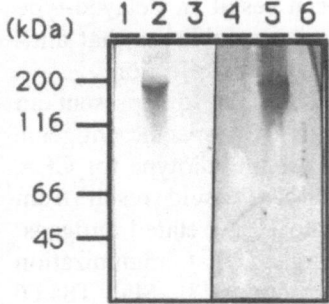

Fig. 10. Demonstration of CEA-specificity of anti-antiidiotypic antiserum (Ab3) by Western Blot. Lanes 1 and 4: Fetuin (200 ng). Lane 2. CEA (10 ng). Lane 5: CEA (200 ng). Lanes 3 and 6: CEA-negative crude lung tumor extract (200 ng). Lanes 1–3 were incubated with anti-CEA antiserum (Dakopatts), 1:10.000 dil. Lanes 4–6 were incubated with Ab3, 1:10 dil

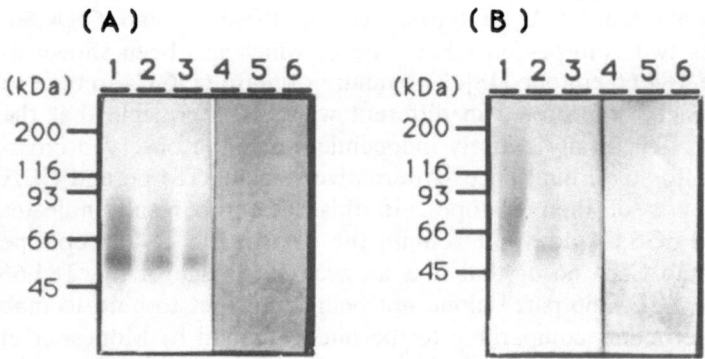

Fig. 11A,B. Crossreactivity of anti-antiidiotypic antiserum (Ab3) with CEA-related antigens by Western Blot. Panel (**A**): Lanes 1 and 6: 100 ng purified NCA. Lanes 2 and 5: 50 ng purified NCA. Lanes 3 and 4: 25 ng purified NCA. Panel (**B**): Lanes 1 and 6: 100 μg tumor extract. Lanes 2 and 5: 10 μg tumor extract. Lanes 3 and 4: 1 μg tumor extract. Immunostaining was identical for both panels. Lanes 1–3 were reacted with anti-CEA (Dakopatts), dil. 1:10.000. Lanes 4–6 were reacted with Ab3, dil. 1:10

different idiotypic cascades [3, 53]. Bhattacharya-Chatterjee et al. [3] investigated the specificity of Ab$_3$ antisera by immunohistochemical means, which is difficult, since CEA-related antigens such as NCA are expressed in various tissues, including colon. On the other hand, in a crude tumor extract the antiserum did not react with glycoproteins that were said to be NCA by molecular weight, indicating specificity, although the presence of CEA-related antigens was not demonstrated immunochemically. The other report showed reaction of the Ab$_3$ antiserum with purified CEA [53]. These authors stated that sequence homology to CEA in the CDRs of their maId M7-625 was observed. However, since paratopes are complex structures defined by conformation, the relatedness to the epitope on the nominal antigen may not be obvious from comparison of primary sequences. Recently, Bentley et al. investigated an idiotype/anti-idiotype system. They reported that idiotypic mimicry of the external image was not obtained, although significant sequence homologies between CDRs and the nominal antigen could be observed [6]. In summary, we can conclude that maId 6G6.C4 can act as a surrogate antigen for CEA in rabbits and may possibly do so in humans. Further work with maId 6G6.C4 is currently in progress in our laboratory and will hopefully provide us with a new reagent for active immunotherapy of CEA-positive gastrointestinal carcinomas.

Future Perspectives with Anti-idiotypic Antibodies

In order to optimize the potential usefulness of anti-idiotype antibodies in human malignancies, it may be important to direct the immune system preferentially towards the paratope structures of a useful maId. In this respect, engineering of murine maIds to chimeric human/mouse antibodies will be important to avoid immune responses that are mainly directed against constant region domains, as has been observed with xenogeneic mabs in humans [13, 57]. Other means of improvement may include the use of cocktails of appropriate maIds that bear images of different CEA-specific epitopes. Buchegger et al. [9] showed that this strategy had a marked effect on the outcome of passive immunotherapy in a tumor xenograft model.

Apart from the immediate medical implications discussed above, internal image anti-idiotypes also can be considered as valuable technical and scientific tools:

1. An idiotype-specific ligand can be used for highly specific affinity chromatography of chimeric idiotype antibody fragments that otherwise are difficult to purify from complex mixtures such as ascites or tissue culture supernatant.
2. The characterization of protein structures is also very important for assessment of structure-function relationships in general, not only in immune complexes. By learning more about the interactions, we might be able to predict structural requirements for efficient protein design. Structural data from crystallization analyses are only available for a limited number of proteins and will most certainly be hard to obtain for highly glycosylated proteins like CEA. On the other hand, copying antigenic determinants onto the antigen-binding "pocket" of an immunoglobulin by sequential immunizations in idiotypic cascades can be expected to circumvent most of these problems. Crystallographic analyses of a number of Fv fragments of immunoglobulins have, with the obvious exception of the antigen-binding regions, revealed the highly con-

served "monotonous" structure of the frameworks of antibody domains. Furthermore, framework key residues influencing the conformation of the CDRs have been identified [12]. For these reasons, antibody molecules are the best understood complex molecular structures so far [25]. Recently, it became possible to predict the structures of Fv fragments with high fidelity from sequence data alone by computer modeling prior to crystallization of the antibody molecules [11]. This indicates that in the very near future X-ray analyses may not be necessary – at least for antibodies – to obtain an accurate model of structure. As a consequence, unknown structures of the immunoglobulin are reduced to the individual antigen-binding regions. Epitopes can be displayed by internal image antibodies, with the effect that noncrystallizable antigens may become functionally accessible after crystallization or by computer modeling of the anti-idiotype. Protein design may ultimately lead to synthetic network antigens the functional significance of which has been previously assessed by means of their respective internal image anti-idiotype antibodies.

Acknowledgment. This work was supported by BMFT grant number 01 GA8712 and by PCO Trust grant number K398. The technical assistance of Birgit Lilienthal and Ursula Fenger is greatly appreciated. I am particularly indebted to Dr. Franz-Josef Gaida for communicating data prior to publication.

References

1. Arakawa F, Kuroki M, Misumi Y, Oikawa S, Nakazato H, Matsuoka Y (1990) Characterization of a cDNA clone encoding a new species of the nonspecific crossreacting antigen (NCA), a member of the CEA gene family. Biochem Biophys Res Commun 166:1063–1071
2. Audette M, Buchegger F, Schreyer M, Mach J-P (1987) Monoclonal antibody against carcinoembryonic antigen (CEA) identifies two new forms of crossreacting antigens of molecular weight 90 000 and 160 000 in normal granulocytes. Mol Immunol 24:1177–1186
3. Battacharya-Chatterjee M, Mukerjee S, Biddle W, Foon KA, Köhler H (1990) Murine monoclonal anti-idiotype antibody as a potential network antigen for human carcinoembryonic antigen. J Immunol 145:2758–2765
4. Beatty JD, Williams LE, Yamauchi D, Morton BA, Hill LR, Beatty BG, Paxton RJ, Merchant B, Shively JE (1990) Presurgical imaging with Indium-labeled anti-CEA for colon cancer staging. Cancer Res [Suppl] 50:922s
5. Beidler CB, Ludwig JR, Cardenas J, Phelps J, Papworth CG, Melcher E, Sierzega M, Myers LJ, Unger BW, Fisher M, David GS, Johnson MJ (1988) Cloning and high level expression of a chimeric antibody with specificity for human carcinoembryonic antigen. J Immunol 141:4953–4960
6. Bentley GA, Boulot G, Riottot MM, Poljak RJ (1990) Three-dimensional structure of an idiotope-anti-idiotope complex. Nature 348:254–257
7. Bona C, Heber-Katz E, Paul WE (1981) Idiotype-antiidiotype regulation. Immunization with a levane-binding myeloma protein leads to appearance of auto-anti-(anti-idiotype) antibodies and to the activation of silent clones. J Exp Med 53:951–967

8. Buchegger F, Haskell CM, Schreyer M, Scazziga BR, Randin S, Carrel S, Mach J-P (1983) Radiolabeled fragments of monoclonal antibodies against carcinoembryonic antigen for localization of human colon carcinoma grafted into nude mice. J Exp Med 158:413–427

9. Buchegger F, Pfister C, Fournier K, Prevel F, Schreyer M, Carrel S, Mach J-P (1989) Ablation of human colon carcinoma in nude mice by 131-I-labeled monoclonal anti-carcinoembryonic antigen antibody F(ab')2 fragments. J Clin Invest 83:1449–1456

10. Buraggi GL, Gasparini M, Seregni E (1991) Immunoszintigraphy of colorectal carcinoma with an anti-CEA monoclonal anti-body: a critical review. Nucl Med Biol 18:45–50

11. Chothia C, Lesk AM, Tramontano A, Levitt M, Smith-Gill SJ, Air G, Sheriff S, Padlan EA, Davies D, Tulip WR, Colman PM, Spinelli S, Alzari PM, Poljak RJ (1989) Conformations of immunoglobulin hypervariable regions. Nature 342:877–883

12. Chothia C, Lesk AM (1987) Canonical structures for the hypervariable regions of immunoglobulins. J Mol Biol 196:901–917

13. Courtenay-Luck NS, Epenetos AA, Moore R, Larche M, Pectasides D, Dhokia B, Ritter MA (1986) Development of primary and secondary immune responses to mouse monoclonal antibodies used in the diagnosis and therapy of malignant neoplasms. Cancer Res 46:6489–6493

14. Dillman RO, Beauregard JC, Sobol RE, Royston I, Bartholomew RM, Hagan PS, Halpern SE (1984) Lack of radioimmunodetection and complications associated with monoclonal anticarcinoembryonic antigen antibody crossreactive with an antigen on circulating cells. Cancer Res 44:2213–2218

15. Drzeniek Z, Lamerz R, Fenger U, Wagener C, Haubeck HD (1991) Identification of membrane antigens in granulocytes and colonic carcinoma cells by a monoclonal antibody specific for biliary glycoprotein, a member of the carcinoembryonic antigen family. Cancer Lett 56:173–179

16. Dunn PL, Johnson CA, Styles JM, Pease SS, Dean CJ (1987) Vaccination with syngeneic monoclonal anti-idiotype protects against a tumour challenge. Immunology 60:181–186

17. Gaida F-J, Fenger U, Wagener C, Neumaier M (1992) A monoclonal anti-idiotypic antibody bearing the image of an epitope specific for the carcinoembryonic antigen. Int J Cancer 52:1–7

18. Gianetti BM, Neumaier M, Wagener C (1986) Monoclonal antibodies reveal repetitive crossreactive and single specific epitopes on the protein moiety of carcinoembryonic antigen. Fres Anal Chem 324:253–254

19. Gold P, Freedman SO (1965) Demonstration of tumor-specific antigens in human colonic carcinomata by immunological tolerance and absorption techniques. J Exp Med 121:439–462

20. Herlyn D, Powe J, Alavi A, Mattis JA, Herlyn M, Ernst C, Vaum R, Koprowski H (1983) Radioimmunodetection of human tumor xenografts by monoclonal antibodies. Cancer Res 43:2731–2735

21. Herlyn D, Wettendorff M, Schmoll E, Iliopoulos D, Schedel I, Dreikhausen U, Raab R, Ross AH, Jaksche H, Scriba M, Koprowski H (1987) Anti-idiotype immunization of cancer patients: modulation of the immune response. Proc Natl Acad Sci USA 84:8055–8059

22. Hinoda Y, Neumaier M, Hefta SA, Drzeniek Z, Wagener C, Shively L, Hefta LJF, Shively JE, Paxton RJ (1988) Molecular cloning of a cDNA coding biliary glycoprotein. I: primary structure of a glycoprotein immunologically crossreactive with carcinoembryonic antigen. Proc Natl Acad Sci USA 85:6959–6963

23. Jakowatz JG, Beatty BG, Vlahos WG, Porudominsky D, Philben VJ, Williams LE, Paxton RJ, Shively JE, Beatty JD (1985) High-specific-activity [111]in-labeled

anticarcinoembryonic antigen monoclonal antibody. Biodistribution and imaging in nude mice bearing human colon cancer xenografts. Cancer Res 45:5700–5706

24. Jerne NK (1974) Towards a network theory of the immune system. Ann Immunol Inst Pasteur 125 C:373–389

25. Kabat EA, Wu TT, Reid-Miller M, Perry HM, Gottesman KS (1987) Sequences of proteins of immunological interest, 4th edn. US Department of Health and Human Services, Washington

26. von Kleist S, Chavanel G, Burtin P (1972) Identification of an antigen from normal human tissue that cross-reacts with the carcinoembryonic antigen. Proc Natl Acad Sci USA 69:2492–2494

27. Kunkel TA (1985) Rapid and efficient site-specific mutagenesis without phenotypic selection. Proc Natl Acad Sci USA 82:488–492

28. Lee VK, Harriott TG, Kuchroo VK (1985) Monoclonal antiidiotypic antibodies related to a murine oncofetal bladder tumor antigen induce specific cell-mediated tumor immunity. Proc Natl Acad Sci USA 82:6286–6290

29. Levy RL, Miller RA (1983) Biological and clinical implications of lymphocyte hybridomas. Annu Rev Med 34:107–116

30. Liu AY, Robinson RR, Hellström KE, Murray ED Jr, Chang CP, Hellström I (1987) Chimeric mouse-human IgG1 antibody that can mediate lysis of cancer cells. Proc Natl Acad Sci 84:3439–3443

31. Maniatis T, Sambrook J, Fritsch EF (1989) Molecular cloning: a laboratory manual, 2nd edn. Cold Spring Harbor Laboratory, Cold Spring Harbor

32. Mellstedt H, Frödin J-E, Biberfeld P, Fagerberg J, Giscombe R, Hernandez A, Masucci G, Li S-L, Steinitz M (1991) Patients treated with a monoclonal antibody (ab1) to the colorectal carcinoma antigen 17-1A develop a cellular response (DTH) to the "internal image of the antigen" (ab2). Int J Cancer 48:344–349

33. Mittelman A, Chen ZJ, Kageshita T, Yang H, Yamada M, Baskind P, Goldberg N, Puccio C, Ahmed T, Arlin Z, Ferrone S (1990) Active specific immunotherapy in patients with melanoma. J Clin Invest 86:2136–2144

34. Monestier M, Debbas ME, Goldenberg DM (1989) Syngeneic antiidiotype monoclonal antibodies to murine anticarcinoembryonic antigen monoclonal antibodies. Cancer Res 49:123–126

35. Nap M, Hammarström ML, Börmer O, Hammarström S, Wagener C, Handt S, Schreyer M, Mach J-P, Buchegger F, von Kleist S, Grunert F, Seguin P, Fuks A, Holm R, Lamerz R (1992) Specificity and affinity of monoclonal antibodies against carcinoembryonic antigen. Cancer Res 52:2323–2339

36. Neumaier M, Fenger U, Wagener C (1985) Monoclonal antibodies for carcinoembryonic antigen (CEA) as a model system: Identification of two novel CEA-related antigens in meconium and colorectal carcinoma tissue by western blots and differential immunoaffinity chromatography. J Immunol 135:3604–3609

37. Neumaier M, Fenger U, Wagener C (1985) Delineation of four carcinoembryonic antigen (CEA) related antigens in normal human plasma bytransblot studies using monoclonal anti-CEA antibodies with different epitope specificities. Mol Immunol 22:1273–1277

38. Neumaier M, Zimmermann W, Shively L, Hinoda Y, Riggs AD, Shively JE (1988) Characterization of a cDNA clone for the nonspecific cross-reacting antigen (NCA) and a comparison of NCA and carcinoembryonic antigen. J Biol Chem 263:3202–3207

39. Neumaier M, Shively L, Chen FS, Gaida F-J, Ilgen C, Paxton RJ, Shively JE, Riggs AD (1990) Cloning of the genes for T84.66, an antibody which has a high

specificity and affinity for the carcinoembryonic antigen (CEA), and expression of chimeric human/mouse T84.66 genes in myeloma and CHO cells. Cancer Res 50:2128–2134

40. Nishimura Y, Yokoyama M, Araki K, Ueda R, Kudo A, Watanabe T (1987) Recombinant human-mouse chimeric monoclonal antibody specific for common acute lymphocytic leukemia antigen. Cancer Res 47:999–1005

41. Oikawa S, Nakazato H, Kosaki G (1987) Primary structure of human carcinoembryonic antigen (CEA) deduced from cDNA sequence. Biochem Biophys Res Commun 142:511–518

42. Queißer W (1990) Chemotherapie gastrointestinaler Tumoren Dtsch Ärztebl 87 (25/26):1478–1480

43. Raychauduri S, Saeki Y, Chen JJ, Iribe H, Fuji H (1987) Tumor-specific idiotype vaccines: II. Analysis of the tumor related network response induced by the tumor and the internal image antigen. J Immunol 139:271–278

44. Riechmann L, Clark M, Waldmann H, Winter G (1988) Reshaping human antibodies for therapy. Nature 332:323–327

45. Sahagan BG, Dorai H, Saltzgaber-Müller J, Toneguzzo F, Guindon CA, Lilly SP, MacDonald KW, Morissey DV, Stone BA, Davis GL, McIntosh PK, Moore GP (1986) A genetically engineered murine/human chimeric antibody retains specificity for human tumor-associated antigen. J Immunol 137:1066–1074

46. Saiki RK, Scharf S, Faloona F, Mullis KB, Horn GT, Erlich HA, Arnheim N (1985) Enzymatic amplification of ß-globin genomic sequences and restriction analysis for diagnosis of sickle cell anemia. Science 230:1350–1354

47. Sayers JR, Schmidt W, Eckstein F (1988) 5'–3' Exonucleases in phosphorothioate-based oligonucleotide-directed mutagenesis. Nucleic Acids Res 16:791–802

48. Sikorska HM (1988) Therapeutic applications of antiidiotypic antibodies. J Biol Response Mod 7:327

49. Shively JE, Beatty JD (1985) CEA-related antigens: molecular biology and clinical significance. Crit Rev Oncol Hematol 2:355–399

50. Sun LK, Curtis P, Rakowicz-Szulczynska E, Ghrayeb J, Chang N, Morrison SL, Koprowski H (1987) Chimeric antibody with human constant region and mouse variable regions directed against carcinoma-associated antigen 17–1A. Proc Natl Acad Sci USA 84:214–218

51. Svenberg T, Hammarström S, Hedin A (1979) Purification and properties of biliary glycoprotein I (BGP-I). Immunochemical relationship to carcino-embryonic antigen. Mol Immunol 16:245–252

52. Tannock IF (1989) Principles of cell proliferation. In: deVita VT Jr, Hellman S, Rosenberg SA (eds) Cancer: principles and practice in oncology 3rd edn. Lippincott, Philadelphia, pp 3–13

53. Tsujisaki M, Imai K, Tokuchi S, Hanzawa Y, Ishida T, Kitagawa H, Hinoda Y, Yachi A (1991) Induction of antigen-specific immune response with the use of anti-idiotypic monoclonal antibodies to carcinoembryonic antigen antibodies. Cancer Res 51:2599–2604

54. Verhoeyen M, Milstein C, Winter G (1988) Reshaping human antibodies: grafting an antilysozyme activity. Science 239:1534–1536

55. Wagener C, Clark BR, Rickard KJ, Shively JE (1983) Antibodies for carcinoembryonic antigen and related antigens as a model system: determination of affinities and specificities of monoclonal antibodies using biotin-labeled antibodies and avidin as precipitating agent in a solution phase immunoassay. J Immunol 130:2302–2307

56. Wagener C, Yang YHJ, Crawford FG, Shively JE (1983) Monoclonal antibodies for carcinoembryonic antigen and related antigens as a model system: a

systematic approach for the determination of epitope specificities of monoclonal antibodies. J Immunol 130:2308–2315
57. Welt S, Divgi CR, Real FX, Yeh SD, Garin-Chesa P, Finstad CL, Sakamoto J, Cohen A, Sigurdson ER, Kemeny N, Carswell EA, Oettgen HF, Old LJ (1990) Quantitative analysis of antibody localization in human metastastatic colon cancer: a phase I study of monoclonal antibody A33. J Clin Oncol 8:1894–1906

Tumor Localization by Immunoscintigraphy: Potential and Limitations

S. Matzku[1] and H. Bihl[2]

[1] Abteilung Immunchemie, Pharmaforschung E. Merck, Frankfurter Straße 250, W-6100 Darmstadt, FRG
[2] Klinik für Nuklearmedizin, Katharinenhospital, W-7000 Stuttgart, FRG

Milestones of Development

It is worth remembering that positive tumor localization with labeled antibodies was one of the founder technologies of nuclear medicine (for review see [1]). However, the enthusiastic approaches in the 1950s were abandoned when it became clear that the target structures detected by the then available polyclonal antitumor antisera were not actually tumor-specific at all, but, for example, directed against ectopic normal components such as peritumoral fibrin [2–4].

This period was followed by the era of oncofetal antigens, i.e. carcino-embryonic antigen (CEA) [5, 6] and alpha-fetoprotein [7, 8], the initial focus being on in vitro diagnosis by the radioimmunoassay technique. Somewhat later, immunoscintigraphy experienced a revival [9, 10] which was nurtured by the introduction of the nude mouse xenograft system [9] and by the usage of affinity-purified antibodies (references in [11]).

The most influential advance with respect to immunoscintigraphy, i.e., monoclonal antibody (MAb) technology, followed shortly thereafter. The impressive progress that was achieved using this methodology rested on two elements: (1) MAbs proved to be highly valuable for the identification of new tumor antigens; (2) at the same time, these MAbs could be used as reproducibly available targeting vehicles. Despite the great potential of the MAb approach, the search for *tumor-specific* antigens was clearly elusive. In a kind of compensatory endeavor many researchers set out to demonstrate that *tumor-associated* antigens which are not exclusively expressed on tumor cells may still represent reasonable targets for diagnostic tumor localization, with the proviso that the antigen density in the target tissue should be high enough and that concurrent antigen expression in normal tissue should be negligible.

The second half of the 1980s was dominated by the search for method-ological improvements with respect to MAb labeling. Iodine nuclides were

Wagener/Neumann (Eds.)
Molecular Diagnostics of Cancer
© Springer-Verlag Berlin Heidelberg 1993

replaced first by ^{111}In chelates [12, 13] and then by ^{99m}Tc, the latter nuclide being either introduced into the disulfide scaffold of the hinge region (Schwartz method; for references see [14]) or also conjugated via a chelate [15]. The most ambitious goal envisaged, but yet not fully accomplished, consists in implementing nuclides or nuclide pairs with the potential for both immunolocalization and radioimmunotherapy (see below) and in using positron-emitting nuclides for immunopositron emission tomography [16].

State of the Art in Immunoscintigraphy

Technetium-99m has clearly paved the way for high-quality immuno-scintigraphic images of many different tumor types (for reviews see, e.g., [17, 18]), diagnostic accuracy being typically above 90% [19]. Interestingly enough, the forerunner of tumor-associated antigens, namely CEA, still seems to compete fairly well with second- and third-generation target molecules. In view of the fact that circulating target antigen does not substantially impair the localizational power of labeled MAbs and that postsurgical monitoring by in vitro assays invariably requires a method for detecting/confirming the source of marker production in vivo, MAb/antigen systems allowing for a combination of in vitro and in vivo diagnosis seem to represent the best candidates for continuing success and broad application. In the absence of back-up by an in vitro monitoring strategy, immuno-localization will be hard pressed to compete with the superb localization achieved by magnetic resonance imaging and computed tomography. Further improvements of radiolabeling techniques and the usage of radio-nuclides and antibody fragments in carefully designed combinations will change little in this respect. More far-reaching expectations are linked to the approach of chimerizing/humanizing MAbs [20, 21]. This will result in a great reduction of human-anti-mouse antibody (HAMA) response [22, 23], thus allowing for multiple applications of the antibody, and in a marked prolongation of the serum half-life of intact antibody [23, 24], which, however, is an ambiguous achievement with respect to imaging contrast.

Future Developments

Novel approaches and methodological advances are expected to emerge from the following:

- Combination of immunoscintigraphy and radioimmunotherapy
- Multimodal targeting strategies involving low-Mr radiopharmaceuticals and bifunctional vehicles
- Shift towards ligand-based tumor localization

Immunoscintigraphy and Radioimmunotherapy

This combination promises substantial contributions to tumor therapy with both labeled and unlabeled antibodies. With respect to serotherapy it is trivial to state that every version of this approach will be based on a detailed knowledge of the kinetics of accumulation of MAbs or MAb conjugates in the target tissue, the biodistribution in the rest of the organism, and the consequences of variations in route or regime (e.g., duration of application, dose variation or MAb modification). All this requires the usage of an approapriately labeled antibody which can be detected by gamma cameras and gamma counters. For illustrations of the potential, the reader is referred to work on the tumor therapy with antibodies against the EGF receptor and related molecules [25–27].

The combination of immunoscintigraphy and radioimmunotherapy (RIT) is almost as classic as immunoscintigraphy per se, since the pioneering attempts at tumor localization by labeled antibodies were readily followed by RIT [4, 28]. In the era of MAbs and MAb-defined target antigens, novel possibilities were elaborated. These were particularly fruitful with respect to leukemias/lymphomas, because a wealth of lymphocyte differentiation antigens was identified with the MAb technology. These served as markers for histologic/cytologic diagnosis and as target antigens for diagnosis and therapy in vivo. Due to the fact that such target antigens are by definition expressed both on normal and malignant lymphocytes, MAb application invariably results in the visualization of normal and malignant lymphatic tissue and yields information as to the distribution of circulating normal and leukemic cells. Figure 1 shows an example from our own work, the sequence of immunoscintigrams being performed 30 min, 12 h, 3 days and 6 days after injection of ^{111}In-labeled MAb HD37 into a patient with B lymphoma processes. The antibody was directed against the CD19 B lymphocyte differentiation antigen [29]. The images illustrate the distribution pattern of the subpopulation of lymphocytes which is positive for the CD19 antigen (i.e., mainly in bone marrow and spleen) and give an indication of the kinetics and the magnitude of labeled antibody accumulation in the lymphoma processes. From such studies a radiation dose estimate may be obtained, provided the therapeutic nuclide is identical to or closely follows the biodistribution of the diagnostic nuclide. Hence, it can be decided on the basis of immunoscintigraphy results whether or not individual patients are amenable to RIT.

Using this strategy, a number of groups have performed RIT studies, success being mainly achieved in the lymphoma indication [30–34] (Table 1). This is most likely due to three factors: (1) the notorious radiosensitivity of lymphomas, (2) the favorable accessibility of leukemic cells and – to some extent – of lymphoma tissue, and (3) the fact that the concomitant elimination of mature normal lymphocytes can be tolerated. Attempts to proceed towards hopefully curative RIT were further stimulated by the possibility of

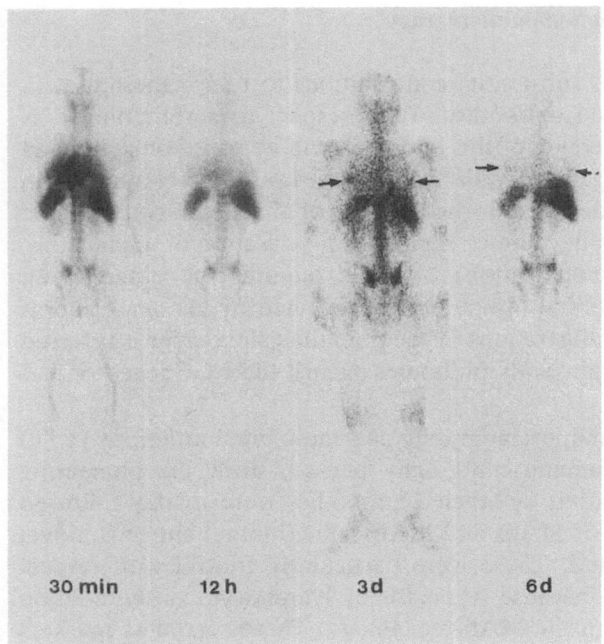

Fig. 1. Immunoscintigrams of a patient with B-lymphoma. The time course analysis of radioactivity distribution performed with ^{111}In-labeled MAb HD37 highlights initially the blood pool, bone marrow, spleen and liver. Lymphoma processes are clearly visualized (*arrows*) after prolonged accumulation time (3 and 6 days). The study was performed in cooperation with B. Dörken and G. Moldenhauer (Heidelberg)

Table 1. Radioimmunotherapy: selected studies

Authors	Tumor entity	No. of patients	Monoclonal antibody	Activity	Response % (PR/CR)
Order et al. 1985 [65]	Hepatocellular carcinoma	66	Antiferritin	20–30 mCi ^{131}I	48 (17/5)
Rosen et al. 1987 [66]	T-lymphoma	5	T101	100–150 mCi ^{131}I	100 (2 + 3/0)
Eary et al. 1990 [34, 67]	Non-Hodgkin's lymphoma	5	MB-1	230–600 mCi ^{131}I	100 (1/4)
Denardo et al. 1990 [31]	B-lymphoma	15	Lym-1	100 mCi/m^2 ^{131}I	100 (8/7)
Bergh et al. 1990 [68]	Astrocytoma	10	MUC 2-63	5–30 mCi ^{90}Y	20 (2/0)

escalating radiation dose and coping with bone marrow toxicity by marrow transplantation [35]. With respect to RIT of solid tumors, available data are scarce and far less promising.

It is fair to state that further improvements of RIT methodology are clearly needed. New labeling procedures may provide improvements with respect to radiation properties and pharmacokinetic as well as biodistributional parameters [36]. The catalogue of suggestions comprises, for example, novel conjugates [37–39] of iodine nuclides (including ^{125}I [40]), ^{90}Y [32, 41–43], and $^{186/188}$Re [19, 44]. Furthermore, humanization of antibodies is expected to be most beneficial, since this will lead to a prolongation of half-life (together with an increased area under the time/concentration curve) and to a dramatic reduction of HAMA induction, the latter point being of paramount importance with respect to repeated MAb application. However, it has to be realized that in the case of chelators conjugated to (humanized) MAbs, antibody induction in patients may be directed against this moiety as well and may prove to be equally deleterious (A. Epenetos, personal communication).

Multimodal Targeting Strategies

The limitations of macromolecule targeting into solid (tumor) tissue have been amply documented [45–47]. In the light of these data, a reduction of the size of the antibody molecule, i.e., down to Fab (50 kDa), Fv (25 kDa [48]), or dab (12 kDa [49, 50]), will have only limited effect, being given that such reduction is invariably accompanied by a dramatic acceleration of plasma clearance [50] and by a loss in affinity in the order of 1–3 logs. Hence, it was desirable to conceive a targeting strategy where the radionuclide is transported by a carrier with Mr below 1 kDa. We now have at our disposal evidence from two different systems:

1. Targeting with bispecific antibodies (bsMAbs). The bsMAb is constructed such that one arm is directed against the tumor-associated antigen while the other arm binds to the radionuclide complex [51–54]. Binding of the bsMAb to the cell surface is equivalent to the introduction of an artificial receptor for the low-Mr radiopharmaceutical. The potential of the approach has been demonstrated by localization studies [54, 55]. According to the findings, the advantages of the bsMAb approach are high localization contrast and short accumulation time. Thus, the approach may be further elaborated towards positron emission tomography, which would allow for a dynamic quantification of uptake in relatively small volume elements. When using appropriate chelators and bsMAbs directed against these [56, 57, 57a], the ^{68}Ga radionuclide, which emits positrons with a 1-h half-life, comes into play. A strategy of this kind may prove to be superior to the use of positron emitters, e.g., ^{124}I [16], for directly labeling the antibody.

2. Combination of MAb, biotin and (strept-)avidin. In this system, [111]In- or ^{99m}Tc-labeled biotin plays the role of the low-Mr radiopharmaceutical, while the avidin component represents the artificial receptor targeted to the tumor cell surface [36, 58, 59]. Preliminary clinical studies with a three-step procedure demonstrate that high contrast and rapid visualization of the tumor may be achieved by careful adjustment of the dosage and the timeframe of application of the components [59, 60].

Targeting of Physiological Receptors with Labeled Ligands

Before considering the introduction of artificial receptors and the use of synthetic ligands, nuclear medicine scientists were already actively engaged in studies that directly addressed the issue of specific ligands to natural receptors. Hence, the recent report on the application of labeled EGF for scintigraphic tumor localization [61] may or may not be considered a sequela to targeting approaches with anti-epidermal growth factor receptor antibodies [25-27]. Another recent example along the lines of targeting bioactive ligands to a receptor that is highly expressed on selected tumor types involves the use of labeled somatostatin or a somatostatin analogue [62, 63]. This work has found enormous interest because it represents yet another approach of linking localizational endeavors to a therapeutic strategy [19, 64] and because it promises the amalgamation of positron emission tomography and tumor targeting by labeled biomolecules [19, 64].

Final Remarks

Biomedical research has thrown up a wealth of new methods whose potential with respect to tumor localization being far from exhausted. However, careful analysis of current evidence reveals that some of the basic facts of physiology, e.g., restrictions on the relative size of compartments or on the transport of macromulecules across tissue barriers, may not be easily overcome. In addition, and most importantly, it has to be accepted that the introduction of immunoscintigraphy into general clinical practice is dependent not on the usefulness of the method in isolation, but on its merits relative to competing methods. In clinical oncology, new diagnostic modalities will meet substantial reservations unless they are accompanied by or lead to new therapeutic approaches. This aspect may prove decisive with respect to the future development of immunolocalization strategies.

References

1. Pressman D (1980) The development and use of radiolabeled antitumor antibodies. Cancer Res 40:2960-2964

2. Dewey WC, Bale WF, Rose RG, Marrack D (1963) Localization of antifibrin antibodies in human tumors. Acta Un Int Cancer 19:187–196
3. Spar IL, Bale WF, Marrack D, Dewey WC, McCardle RJ, Harper PV (1967) 131I-labeled antibodies to human fibrin. Cancer 20:865–870
4. Bale WF, Contreras MA, Grady ED (1980) Factors influencing localization of labeled antibodies to tumors. Cancer Res 40:2965–2972
5. Gold P, Freedman SO (1965) Specific carcinoembryonic antigens of the human digestive system. J Exp Med 122:467–481
6. Gold P, Shuster J (1980) Historical development and potential uses of tumor antigens as markers of human cancer growth. Cancer Res 40:2973–2976
7. Ruoslahti E, Seppälä M (1971) Studies of carcinofetal protein: III. Demonstration of alpha-fetoprotein in serum of healthy adults. Int J Cancer 8:374–379
8. Kim EE, Deland FH, Nelson MO, Bennett S, Simmons G, Alpert E, Goldenberg DM (1980) Radioimmunodetection of cancer with radiolabeled antibodies to alpha-fetoprotein. Cancer Res 40:3008–3012
9. Mach J-P, Carrel S, Merenda C (1974) In vivo localization of radiolabeled antibodies to carcinoembryonic antigen in human colon carcinoma grafted into nude mice. Nature 248:704–706
10. Goldenberg DM, DeLand F, Kim E, Bennett S, Primus FJ, van Nagell JR, Estes N, DeSimone P, Rayburn P (1978) Use of radiolabeled antibodies to carcinoembryonic antigen for the detection and localization of diverse cancers by external photoscanning. N Engl J Med 298:1384–1388
11. Mach J-P, Buchegger F, Forni M, Ritschard J, Berche C, Lumbroso J-D, Schreyer M, Girardet C, Accolla RS, Carrel S (1981) Use of radiolabelled monoclonal anti-CEA antibodies for the detection of human carcinomas by external photoscanning and tomoscintigraphy. Immunol Today 2:239–249
12. Hnatowich DJ, Childs RL, Lanteigne D, Najafi A (1983) The preparation of DTPA-coupled antibodies radiolabeled with metallic radionuclides: an improved method. J Immunol Methods 65:147–157
13. Meares CF, McCall MJ, Reardan DT, Goodwin DA, Diamanti CI, McTigue M (1984) Conjugation of antibodies with bifunctional chelating agents: isothiocyanate and bromoacetamide reagents, methods of analysis, and subsequent addition of metal ions. Anal Biochem 142:68–78
14. Mather SJ, Ellison D (1990) Reduction-mediated technetium-99m labeling of monoclonal antibodies. J Nucl Med 31:692–697
15. Fritzberg AR (1987) Advances in the 99mTc-labeling of antibodies. Nucl Med 26:7–12
16. Wilson CB, Snook DE, Dhokia B, Taylor CVJ, Watson IA, Lammertsma AA, Lambrecht R, Waxman J, Jones T, Epenetos AA (1991) Quantitative measurement of monoclonal antibody distribution and blood flow using positron emission tomography and [124]iodine in patients with breast cancer. Int J Cancer 47:344–347
17. Crowther ME, Britton KE, Granowska M, Shepherd JH (1989) Monoclonal antibodies and their usefulness in epithelial ovarian cancer. Br J Obstet Gynaecol 96:516–521
18. Baum RP, Lorenz M, Senekowitsch R, Chatal JF, Saccavini JC, Hottenrott C, et al. (1988) Clinical experience in cancer diagnosis with radiolabeled monoclonal antibodies in 200 patients and initial attempts at radioimmunotherapy. In: Srivastava SC (ed) Radiolabeled monoclonal antibodies for imaging and therapy. Plenum, New York, pp 613–651
19. Wagner HN Jr (1991) Molecular medicine: from science to service. J Nucl Med 32:11N–23N
20. Sun LK, Curtis P, Rakowicz-Szulczynska E, Ghrayeb J, Chang N, Morrison SL, Koprowski H (1987) Chimeric antibody with human constant regions and mouse

variable regions directed against carcinoma-associated antigen 17-1A. Proc Natl Acad Sci USA 84:214–218

21. Riechmann L, Clark M, Waldmann H, Winter G (1988) Reshaping human antibodies for therapy. Nature 332:323–327

22. Hale G, Dyer MJ, Clark MR, Phillips JM, Marcus R, Riechmann L, Winter G, Waldmann H (1988) Remission induction in non-Hodgkin lymphoma with reshaped human monoclonal antibody CAMPATH-1H. Lancet 2:1394–1399

23. Meredith RF, LoBuglio AF, Plott WE, Orr RA, Brezovich IA, Russell CD, Harvey EB, Yester MV, Wagner AJ, Spencer SA, Wheeler RH, Saleh MN, Rogers KJ, Polansky A, Salter MM, Khazaeli MB (1991) Pharmacokinetics, immune response, and biodistribution of iodine-131-labeled chimeric mouse/ human IgG1,k 17-1A monoclonal antibody. J Nucl Med 32:1162–1168

24. LoBuglio AF, Wheeler RH, Trang J, Haynes A, Rogers K, Harvey EB, Sun L, Ghrayeb J, Khazaeli MB (1989) Mouse/human chimeric monoclonal antibody in man: Kinetics and immune response. Proc Natl Acad Sci USA 86:4220–4224

25. Saga T, Endo K, Akiyama T, Sakahara H, Koizumi M, Watanabe Y, Nakai T, Hosono M, Yamamoto T, Toyoshima K, Konishi J (1991) Scintigraphic detection of overexpressed c-erbB-2 protooncogene products by a class-switched murine anti-c-erbB-2 protein monoclonal antibody. Cancer Res 51:990–994

26. Divgi CR, Welt S, Kris M, Real FX, Yeh SDJ, Gralla R, Merchant B, Schweighart S, Unger M, Larson SM, Mendelsohn J (1991) Phase I and imaging trial of indium 111-labeled anti-epidermal growth factor receptor monoclonal antibody 225 in patients with squamous cell lung carcinoma. J Natl Cancer Inst 83:97–104

27. Kalofonos HP, Pawlikowska TR, Hemingway A, Courtenay-Luck N, Dhokia B, Snook D, Sivolapenko GB, Hooker GR, McKenzie CG, Lavender PJ, Thomas DGT, Epenetos AA (1989) Antibody guided diagnosis and therapy of brain gliomas using radiolabeled monoclonal antibodies against epidermal growth factor receptor and placental alkaline phosphatase. J Nucl Med 30:1636–1645

28. Bale WF, Spar IL, Goodland RL (1960) Experimental radiation therapy of tumors with ^{131}I carrying antibodies to fibrin. Cancer Res 20:1488–1494

29. Pezzutto A, Dörken B, Rabinovitch PS, Ledbetter JA, Moldenhauer G, Clark EA (1987) CD19 monoclonal antibody HD37 inhibits anti-immunoglobulin-induced B cell activation and proliferation. J Immunol 138:2793–2799

30. Badger CC, Bernstein ID (1986) Prospects for monoclonal antibody therapy of leukemia and lymphoma. Cancer 58:584–589

31. DeNardo GL, DeNardo SJ, O'Grady LF, Levy NB, Adams GP, Mills SL (1990) Fractionated radioimmunotherapy of B-cell malignancies with ^{131}I-Lym-1. Cancer Res 50 [Suppl]:1014s–1016s

32. Parker BA, Vassos AB, Halpern SE, Miller RA, Hupf H, Amox DG, Simoni JL, Starr RJ, Green MR, Royston I (1990) Radioimmunotherapy of human B-cell lymphoma with ^{90}Y-conjugated antiidiotype monoclonal antibody. Cancer Res 50 [Suppl]:1022s–1028s

33. Goldenberg DM, Horowitz JA, Sharkey RM, Hall TC, Murthy S, Goldenberg H, Lee RE, Stein R, Siegel JA, Izon DO, Burger K, Swayne LC, Belisle E, Hansen HJ, Pinsky CM (1991) Targeting, dosimetry, and radioimmunotherapy of B-cell lymphomas with iodine-131-labeled LL2 monoclonal antibody. J Clin Oncol 9:548–564

34. Eary JF, Press OW, Badger CC, Durack LD, Richter KY, Addison SJ, Krohn KA, Fisher DR, Porter BA, Williams DL, Martin PJ, Appelbaum FR, Levy R, Brown SL, Miller RA, Nelp WB, Bernstein ID (1990) Imaging and treatment of B-cell lymphoma. J Nucl Med 31:1257–1268

35. Morton BA, Beatty BG, Mison AP, Wanek PM, Beatty JD (1990) Role of bone marrow transplantation in ^{90}Y antibody therapy of colon cancer xenografts in nude mice. Cancer Res 50 [Suppl]:1008s–1010s
36. Goodwin DA (1988) Pharmacokinetics and antibodies. J Nucl Med 28: 1358–1362
37. Wilbur DS, Hadley SW, Hylarides MD, Abrams PG, Beaumier PA, Morgan AC, Reno JM, Fritzberg AR (1989) Development of a stable radioiodinating reagent to label monoclonal antibodies for radiotherapy of cancer. J Nucl Med 30:216–226
38. Badger CC, Wilbur DS, Hadley SW, Fritzberg AR, Bernstein ID (1990) Biodistribution of p-iodobenzoyl (PIP) labeled antibodies in a murine lymphoma model. Nucl Med Biol 17:381–387
39. Zalutsky MR, Narula AS (1988) Radiohalogenation of a monoclonal antibody using an N-succinimidyl 3-(tri-n-butylstannyl) benzoate intermediate. Cancer Res 48:1446–1450
40. Brady LW, Markoe AM, Woo DV, Amendola BE, Karlsson UL, Rackover M, Koprowski H, Steplewski Z, Peyster RG (1990) Iodine-125-labeled anti-epidermal growth factor receptor-425 in the treatment of glioblastoma multiforme. Ant Immun Radiopharm 24:151–160
41. Sharkey RM, Kaltovich FA, Shih LB, Fand I, Govelitz G, Goldenberg DM (1988) Radioimmunotherapy of human colonic cancer xenografts with 90Y-labeled monoclonal antibodies to Carcinoembryonic antigen. Cancer Res 48:3270–3275
42. Esteban JM, Hyams DM, Beatty BG, Wanek P, Beatty JD (1989) Effect of yttrium-90-labeled anti-carcinoembryonic antigen monoclonal antibody on the morphology and phenotype of human tumors grown as peritoneal carcinomatosis in athymic mice. Cancer 63:1343–1352
43. Deshpande SV, DeNardo SJ, Kukis DL, Moi MK, McCall MJ, DeNardo GL, Meares CF (1990) Yttrium-90-labeled monoclonal antibody for therapy: labeling by a new macrocyclic bifunctional chelating agent. J Nucl Med 31:473–479
44. Vanderheyden J-L, Su F-M, Venkatesan P, Beaumier P, Bugaj J, Fritzberg AR (1990) The chemistry of rhenium-186 labeled antibodies and F(ab')2 fragments for RIT in animals and man. J Nucl Med 31:823
45. Cobb LM (1989) Intratumour factors influencing the access of antibody to tumour cells. Cancer Immunol Immunother 28:235–240
46. Jain RK (1990) Physiological barriers to delivery of monoclonal antibodies and other macromolecules in tumors. Cancer Res 50 [Suppl]:814s–819s
47. Fujimori K, Covell DG, Fletcher JE, Weinstein JN (1989) Modeling analysis of the global and microscopic distribution of immunoglobulin G, F(ab')$_2$, and Fab in tumors. Cancer Res 49:5656–5663
48. Skerra A, Pfitzinger I, Plückthun A (1991) The functional expression of antibody F$_v$ fragments in Escherichia coli: improved vectors and a generally applicable purification technique. Biotechnology 9:273–278
49. Ward ES, Güssow D, Griffiths AD, Jones PT, Winter G (1989) Binding activities of a repertoire of single immunoglobulin variable domains secreted from Escherichia coli. Nature 341:544–546
50. Larson SM (1990) Improved tumor targeting with radiolabeled, recombinant, single-chain, antigenbinding protein. J Natl Cancer Inst 82:1173–1175
51. Reardan DT, Meares CF, Goodwin DA (1985) Antibody against metal chelates. Nature 316:265–268
52. Goodwin DA, Meares CF, McCall MJ, McTigue M, Chaovapong W (1988) Pre-targeted immunoscintigraphy of murine tumors with indium-111-labeled bifunctional haptens. J Nucl Med 29:226–234

53. Le Doussal J-M, Martin M, Gautherot E, Delaage M, Barbet J (1989) In vitro and in vivo targeting of radiolabeled monovalent and divalent haptens with dual specificity monoclonal antibody conjugates: enhanced divalent hapten affinity for cell-bound antibody conjugate. J Nucl Med 30:1358–1366

54. Le Doussal J-M, Gautherot E, Martin M, Barbet J, Delaage M (1991) Enhanced in vivo targeting of an asymmetric bivalent hapten to double-antigen-positive mouse B cells with monoclonal antibody conjugate cocktails. J Immunol 146:169–175

55. Gridley DS, Ewart KL, Cao JD, Stickney DR (1991) Hyperthermia enhances localization of ^{111}In-labeled hapten to bifunctional antibody in human colon tumor xenografts. Cancer Res 51:1515–1520

56. Schuhmacher J, Matys R, Hauser H, Maier-Borst W, Matzku S (1986) Labeling of monoclonal antibodies with a 67Ga-phenolic aminocarboxylic acid chelate: I. Chemistry and labeling technique. Eur J Nucl Med 12:197–404

57. Matzku S, Moldenhauer G, Kalthoff H, Canevari S, Colnaghi MI, Schuhmacher J, Bihl H (1990) Antibody transport and internalization into tumors. Br J Cancer 62 (S.X):1–5

57a. Schuhmacher J, Klivenyi G, Hull WE, Matys R, Hauser H, Kalthoff H, Schmiegel W, Maier-Borst W, Matzku S (1992) A bifunctional HBED-derivative for labeling of antibodies with Ga, In and Fe. Comparative biodistribution in mice bearing antibody internalizing and non-internalizing tumors. Nucl Med Biol (in press)

58. Hnatowich DJ, Virzi F, Ruschkowski M (1987) Investigations of avidin and biotin for imaging applications. J Nucl Med 28:1294–1302

59. Paganelli G, Pervez S, Siccardi AG, Rowlinson G, Deleide G, Chiolerio F, Malcovati M, Scassellati GA, Epenetos AA (1990) Intraperitoneal radio-localization of tumors pre-targeted by biotinylated monoclonal antibodies. Int J Cancer 45:1184–1189

60. Paganelli G, Magnani P, Rossetti C, Zito F, Belloni C, Pasini A, Sassi I, Sanvito F, Siccardi AG, Fazio F (1990) Antibody guided tumor detection in CEA positive patients using the avidin-biotin system. J Nucl Med 31:735 (abstr)

61. Schatten C, Pateisky N, Vavra N, Ehrenbock P, Angelberger P, Sivolapenko G, Epenetos A (1991) Lymphoscintigraphy with ^{123}I-labelled epidermal growth factor. Lancet 337:395–396

62. Lamberts SWJ, Bakker WH, Reubi JC, Krenning EP (1990) Somatostatin receptor imaging in the localization of endocrine tumors. N Engl J Med 323:1246–1249

63. Bakker WH, Krenning EP, Breeman WA, Kooij PP, Reubi JC, Koper JW, deJong M, Lameris JS, Vieers TJ, Lamberts SW (1991) The in vivo use of a radioiodinated somatostatin analogue: dynamics, metabolism, and binding to somatostatin receptor-positive tumors in man. J Nucl Med 32:1184–1188

64. Larson SM (1991) Receptors on tumors studies with radionuclide scintigraphy. J Nucl Med 32:1189–1191

65. Order SE, Stillwagon GB, Klein JL, Leichner PK, Siegelman SS, Fishman EK, Ettinger DS, Haulk T, Kopher K, Finney K, Surdyke M, Self S, Leibel S (1985) Iodine 131 antiferritin, a new treatment modality in hepatoma: a Ratiation Therapy Oncology Group Study J Clin Oncol 3:1573–1582

66. Rosen ST, Zimmer AM, Goldman-Leikin R, et al. (1987) Radio-immunodetection and radioimmunotheraphy of cutaneous T-cell lymphomas using an I-131-labeled monoclonal antibody: an Illinois Cancer Council study. J Clin Oncol 5:562–573

67. Press OW, Eary JF et al. (1989) Treatment of refractory NHLs with radiolabeled MB-1 (anti-CD37) antibody. J Clin Oncol 7:1027–1038

68. Bergh J, Nilsson S, Liljedahl C, Sivolapenko G, Maripuu E, Stavrou D, Epenetos A (1990) In vivo imaging and treatment of human brain tumours utilizing the radiolabelled monoclonal antibody MUC 2-63 Anticancer Res 10:655-660

Subject Index

abnormalities, chromosomal 47
acute lymphoblastic leukemia 77
adenocarcinoma, pancreatic 176
adenomatous polyposis 19
adenovirus E1A 30
adherence 88
adhesion, cell 109
  molecule, cell 102
    cell-cell (CAMs) 119
    intracellular 101
    neural cell 117, 119
adjuvant therapy, systemic 172
allele-specific oligodeoxynucleotides 9
alpha-2,8-linked neuraminic acid 121
alternative splicing 119
amplification in vitro 1
  ERB-B2 169, 172
  multiplex, of several regions 7
antibodies, anti-idiotypic 198
  human anti-mouse 189
antibody fragments 192
antigen, carcinoembryonic 107
  early progression 99
  human carcinoembryonic 188
  nonspecific cross-reacting (NCA) 108
  tumor-associated 207
anti-oncogenes 27
antisense oligonucleotides 182
autocrine growth factor secretion 153
  loops 185

basaliomas 87
bcr/abl fusion 13
biliary glycoprotein 108
bispecific antibodies 211
bone marrow transplantation, CML 82

breakpoint 21
  cluster region 80

cancer, bladder 57
  breast 87, 140
  colon 87
  pancreas 57
carcinoembryonic antigen (CEA) 107, 187
  gene 188
  specific antibodies, cocktails 112
carcinogenesis, colorectal 21
  multistep 27
carcinoma 43, 187
  colorectal 187
  genital 65
  hepatocellular 32
carry-over 6
cathepsin D 141
CD19 B lymphocyte differentiation antigen 209
CD44, lymphocyte glycoprotein 89
  splice variant 89
  variant 91
cell adhesion 109
  molecule (CAMs) 102, 119
c-erb-B2 (HER-2) 158
chemical mismatch cleavage (HOT technique) 58
"chemical tags" 1
chimeric immunoglobulins 192
chromogranin A 124
chromogranins/secretogranins 117
chromosome, Philadelphia 17
cleavage, chemical mismatch 58
CML after bone marrow transplantation 82

myc   179
cocktails of CEA-specific antibodies
    112
clonospecific probes   78
    TCR δ   81
clusters of differentiation (CD)66   111
    (CD)67   111
coamplification   11
colorectal carcinomas   187
cooperation of two tumor suppressor
    genes   19
confocal laser scan microscopy   136
cutaneous melanoma progression   97
cytofluorometry   136
cycloheximide   180

deleted in colonic cancer (DCC)   102,
    110
denaturating gradient gel electrophore-
    sis   9
"deadhesion"   88
degradation, matrix   88
depletion, T-cell   82
differential splicing   109
disease, graft-versus-host   82
    minimal residual   78
DNA fingerprinting   41
    fingerprints   43
    sequence homology   67
    sequences, HPV6   69
        HPV   11
    sequencing   10
    template   1
double minutes   41

electrophoresis, denaturating gradient
    gel   9
    temperature gradient gel   9
endothelial recognition   88
enolase, neuro-specific   117, 124
epidermal growth factor (EGF)   154,
    155, 175
    receptor (EGFR)   41, 156, 157,
        171
        extracellular domain   158
        gene   45
epidermodysplasia verruciformis   65
ERB-B2 amplification   169, 172
erb-B2/c-neu, receptor   175
erb-B3   175
error rate   4
erythroleukemia, Friend virus-induced
    30
estrogen receptor   172

exonuclease, 5'-3'   4
extravasation   88

familial neuroma   19
farnesylation   53, 60
FGF   154, 155
fingerprints, DNA   43
fragment of immunoglobulins, Fv   201
fusion, *bcr/abl*   13

GAP, p120   55
    proteins   55
gene, CEA   189
    dosage effect   19, 21
    EGFR   45
    nm23   54
    pleiotropic (metastogene)   93
    retinoblastoma   28
    suppressor   161
glial tumors   151
glioblastomas   151
gliomas   41, 43, 151
    genetics   161
glycoprotein, biliary   108
    pregnancy-specific   108
glycosyl-phosphatidylinositol (GPI)
    133
        anchor   109
graft-versus-host disease (GVHD)   82
growth factor, epidermal   175
    receptors, regulation by cytokines
        178
    transforming   175
GTPase cycle   54

haplotype, MHC   88
HER-2/Neu/c-ERB-B2   169
"heterospecificity" of p53 mutations
    32
heterozygosity, loss of   18, 20, 161
hinge region   192
homology sequence, DNA   67
    units of immunoglobulins   120
HPV 6, DNA sequence   69
HPV 11, DNA sequences   69
HPV 41, DNA   69
HPV infection, genital   73
human anti-mouse antibodies
    (HAMA)   189
human carcinoembryonic antigen
    (CEA)   188
hybridization, dot/slot   72
    In-gel   41

in situ   70
    filter   70
    nucleic acid   69
    reverse blot   71
    Southern blot   71
hypomethylation of tumor DNA   47

IGFs   154
immune therapy, passive   192
immunofluorescence, double-color   79
immunoglobulins, chimeric   192
    chimerized   187
    Fv fragments of   201
    homology units   120
    superfamily   108, 121
    supergene family   102
    xenogeneic   189
immunolocalization   107
immunoperoxidase   176
immunoscintigraphy   209
immunotherapy, active   187
    passive specific   187
inhibitor, protein C   129
integrin VLA-5   88
intracellular adhesion molecule 1
    (ICAM-1)   101, 103
invasion, tumor   141, 158
isopsoralens   7

Karyotype analysis   43
Klenow fragments of *E. coli*
    polymerase I   3
"Knudson model"   17

leukemia   77
    acute lymphoblastic   77
    myeloid   57
LFA-1   103
Li-Fraumeni codon 248 mutants   35
    syndrome   32
ligase chain reaction (LCR)   12
ligation chain reaction   2
localization, tumor   209
LOH (loss of heterozygosity)   18, 20, 30
lovastatin   59
lymphocyte differentiation antigen,
    CD19 B   209
    glycoprotein   CD44   89

MAb, chimerizing/humanizing   208
matrix degradation   88
mechanism, one-hit   17

two-hit   17
medulloblastomas   43
melanoma, cutaneous, progression   97
    primary   98
meningioma   18, 19
metastatic potential   98
metastasis   141
    localization   172
metastogene   91
MHC haplotype   88
migration   158
minimal residual disease   78
minisatellite structures   42
missense point mutations   31
molecule, uPA-R   133
mutagenesis, site directed   193
    dominant-negative   34
    dominant-positive   35
    missense point   31
    null   34
myeloid leukemia   57
    lineage, surface markers   111

neural cell adhesion molecules
    (NCAM)   117, 119
neuraminic acid, alpha-2, 8-linked   121
neurofibromatosis 1 (NF1)   19, 55
    2 (NF2)   19
neuroma, familial   19
    sporadic acoustic   19
neuron-specific enolase   124
neurotransmitters   118
nevi, dysplastic   98
nexin   129
NGF   154
Nick translation   69
nm23 gene   54
nonspecific cross-reacting antigen
    (NCA)   108, 199
nucleic acid, antisense   59
    hybridization   69

oligodeoxynucleotides, allele-specific   9
    antisense   182
    probes   42
oncogene   27
    anti-   27
    HER-2/Neu/c-ERB-B2   169
    recessive   27

p53   28, 163
    antibodies   37
    metabolic stabilization   29

mutant allele, dominance   31
mutations, heterospecificity   32
wild-type   29
p120GAP   55
PAI-1   136, 141
PAI-2   136
palmitylation   53
pancreatitis, chronic   177
papillomavirus genome   66
  group-specific virus   68
  human   65
passive immunotherapy, specific   187
PDGF-R   155
PCR (polymerase chain reaction)   1,
  72, 77, 111, 194
  allele-specific   8
  cycling parameters   5
  diagnostic applications   12
  post-, sterilization   7
  pre-, sterilization   6
  products, analysis   7
    ethidium staining   8
  quantitative   10
  sequencing   10
  specificity   6
PDGF   154, 155
phenotype, metastatic   87
Philadelphia (Ph) translocation   80
  chromosome   17
physophilin   119
plasma clearance   211
plasminogen activator, inhibitors   129
    tissue-type   129
    urokinase-type   129
pleiotropic gene (metastogene)   93
  regulator   91
point mutations, detection   7
polymerase, Klenow fragments of
  E. coli   3
  Pfu DNA   4
  Taq   3, 4
  Vent DNA   4
polymorphism, restriction fragment
  length – (RFLP)   58
  sequence-specific conformational   58
positron emission tomography   211, 212
pRB   163
pregnancy-specific glycoprotein   108
primers   1, 3, 111
  consensus   72
  mismatched   3
probes, clonospecific   78
progesterone receptor   172
progression antigens, early   99
  markers, late   99
  tumor   87

protein C inhibitor   129
  kinase C   56
protein, GAP   55
  heat shock   33
  serpin superfamily   129
  ras-related   58

radioimmunotherapy   209
random-primer synthesis   69
ras family   53
  related proteins   58
R-band regions   42
receptor, erb-B2/c-neu   175
  estrogen   172
  progesterone   172
von Recklinghausen's disease   19
recognition, endothelial   88
regulation of growth factor receptors by
  cytokines   178
regulator, pleiotropic   91
replication, self-sustained sequence –
  2, 12
restriction fragment length polymorphisms (RFLP)   58
retinoblastoma   18, 19
  gene   28

sandwich capture technique   10
scan microscopy, confocal laser
  136
sequence-specific conformational polymorphism (SSCP)   58
sequencing, DNA   10
  PCR   10
serotherapy   209
serpin superfamily of proteins   129
somatostatin   212
specificity of PCR   6
spheroids, multicellular   159
splice variant of CD44   89
splicing, alternative   119
  differential   109
sporadic acoustic neuroma   19
sterilization, post-PCR   7
  pre-PCR   6
Stoffel fragment   4
streptavidin   212
structures, minisatellite   42
suramin   153, 159
SV40 T antigen 20, 24, 28
surface markers for the myeloid
  lineage   111
synaptic vesicles   117, 118
synaptophysin   118, 122

synthesis, random-primer   69
  of single-stranded RNA probes   69

T-cell depletion   82
TCR δ rearrangements   78
Technetium-99m   208
temperature gradient gel
  electrophoresis   9
template, DNA   1
TGF α   155
therapy, adjuvant systemic   170
thermocyclers   7
TNF-α   175, 178
TNF-β   175, 180
transcriptase, reverse   12
transcription, reverse   2
transmembrane domain   109
tumor DNA, hypomethylation   47
  invasion   141, 158
  localization   209
  necrosis factor alpha   175, 178
  progression   87
    genes   27
    cooperation   19

tumors, glial   151
type 2 fimbraiae   110
tyrphostins   182

uPA, assessment   135
uPA-receptor   137
uPA-R molecule   133
uracil-*N*-glycosylase   7
urokinase-type plasminogen activator
  129

variant CD44   91
vesicles, sympatic   117, 118
*v-sis*161

Wilm's tumor   19

xenogeneic immunoglobulins   189